The Sleeping Beauties

Dr Suzanne O'Sullivan has been a consultant in [illegible] gy since 2004, first working at t[illegible] ow as a consultant in clinical [illegible] at the National Hospital for [illegible] e specializes in the investiga[illegible] has an active interest in psych[illegible] first book, *It's All in Your Head*, won both the Wellcome Book Prize and the Royal Society of Biology Book Prize and her critically acclaimed *Brainstorm* was published in 2018.

'O'Sullivan doesn't offer easy answers. She just shows us, with wonderful compassion and the minimum of judgment, the ways in which people across the world have manifested symptoms that have helped them through – or beyond – painful situations.'

Helen Brown, *Daily Telegraph*

'O'Sullivan travels the world collecting fascinating stories of culture-bound syndromes, which she relays with nuance and sensitivity.' Alice Robb, *New Statesman*

'I finished it feeling thrillingly unsettled, and wishing there was more.' James McConnachie, *Sunday Times*

'Illuminating and often challenging . . . By making social problems visible on the body, O'Sullivan believes, these conditions allow voiceless people to make themselves heard. Perhaps this eloquent and convincing book will be the start of making people in authority listen, make change and help.' Katy Guest, *Guardian*

'The stories are remarkable. But no less remarkable is O'Sullivan's revelation of the way we all absorb cultural expectations of illness and reject or exhibit symptoms in response.'

Wendy Moore, *Literary Review*

‘O’Sullivan’s beautifully written book interweaves the stories of those afflicted in this way around the world in a travelogue of illness that is ultimately a travelogue of our own irrational, suggestible minds . . . It is a measure of how effective O’Sullivan is at describing the dilemmas and difficulties of treating psychosomatic conditions that, by the end, a visit to a witch doctor begins to feel like the most sensible medical intervention in the book.’ Tom Whipple, *The Times*

‘In this fascinating book, O’Sullivan makes a case for empathy.’ *iNews*

‘Each case study peels back the rigid framework of modern medicine and demands that we reframe our understanding of what is and isn’t illness. This is a progressive book that doesn’t hold back on criticising the dogged diagnostic obsessions of Western medicine.’ Lucy Kehoe, *Geographical*

‘No one doubts that there is something genuinely wrong with these children, yet medicine cannot locate it. O’Sullivan tours the clusters to see if she can do any better.’ *Strong Words*

‘In my view the best science writer around – a true descendant of Oliver Sacks.’ Sathnam Sanghera, author of *Empireland*

The Sleeping Beauties

And Other Stories of Mystery Illness

SUZANNE O'SULLIVAN

PICADOR

First published 2021 by Picador

This paperback edition first published 2022 by Picador
an imprint of Pan Macmillan
The Smithson, 6 Briset Street, London EC1M 5NR
EU representative: Macmillan Publishers Ireland Ltd, 1st Floor,
The Liffey Trust Centre, 117–126 Sheriff Street Upper,
Dublin 1, D01 YC43
Associated companies throughout the world
www.panmacmillan.com

ISBN 978-1-5290-1057-2

1 3 5 7 9 8 6 4 2

A CIP catalogue record for this book is available from the British Library.

Typeset in Fournier MT Std by Palimpsest Book Production Ltd, Falkirk, Stirlingshire
Printed and bound by CPI Group (UK) Ltd, Croydon, CR0 4YY

Contents

To try to understand the experience of another
it is necessary to dismantle the world
as seen from one's own place within it and
to reassemble it as seen from his.

John Berger, *A Seventh Man*, 1975

Preface: The Mystery Illness

Mystery: *Anything that is kept secret or remains unexplained or unknown.*

I first heard about it from a news website in late 2017. The article, headlined *Sweden's Mystery Illness*, told the story of Sophie, a nine-year-old girl who had fallen into a lifeless, unresponsive state more than a year before. Sophie couldn't move or communicate. She couldn't eat or even open her eyes. In fact, for a very long time, all she had done was to lie completely still, showing no indication of knowing day from night.

The article included a picture of Sophie. It showed her wrapped in a pink blanket. Pinned to the yellow striped wallpaper behind her were a child's drawings – hers, probably, from an earlier time. She was not in hospital. This was her own bed, in her own bedroom. Despite her unresponsive state, medical tests suggested that her brain was healthy. Rather than explaining the coma, her scans said she was not in one at all. With nothing to actively treat, the doctors had sent her home to be cared for by her family, and there she lay, many months later, getting neither better nor worse.

Although the headline made it sound as if Sophie's illness was a complete enigma, the body of the article suggested that the cause was not entirely unknown. Sophie had entered Sweden as an asylum seeker. In her home country of Russia, her family had been subject to persecution by the local mafia. Sophie had seen her mother beaten and her father seized by police. Her illness had started very soon after she and her family had fled Russia and entered Sweden. With good reason, her doctors assumed that the illness had a psychological cause.

As a neurologist, I am familiar with the power of the mind over the body – more than most doctors, perhaps. I regularly see patients lose consciousness through a psychological mechanism rather than as a disease process. I would not consider this phenomenon to be at all rare, or even unusual. At least a quarter of those referred to me with seizures, many of whom believe they have epilepsy, prove to be suffering from dissociative, or psychosomatic, seizures. Those high numbers are not specific to my clinical practice. Up to a third of people attending any neurology clinic have a medical complaint that is likely to be psychosomatic in nature – meaning *real* physical symptoms that are disabling, but which are not due to disease and are understood to have a psychological or behavioural cause. Paralysis, blindness, headache, dizziness, coma, tremor, or any other symptom or disability one can imagine, has the potential to be psychosomatic. And, of course, it's not just a neurological phenomenon; any organ in the body can be affected and almost any symptom can be generated in this way – skin rashes, breathlessness, chest pain, palpitations, bladder problems, diarrhoea, stomach cramps, and on and on.

Despite the ubiquitous nature of these disorders, many people still doubt them, considering them somehow less 'real' than other sorts of medical problems. I admit, I struggle to see where this underestimation comes from. I am aware of all the ways in which my own body speaks for me, often unbidden.

My posture changes with my mood. Poorly controlled facial expressions inadvertently reveal my opinion to others, even when I don't intend it. It doesn't seem a stretch, therefore, to suggest this embodiment of a person's inner world has the potential to extend into illness. That the body is the mouthpiece of the mind seems self-evident to me, but I have the sense that not everybody feels the connection between bodily changes and the contents of their thoughts as vividly as I do. So, when a child becomes catatonic in the context of stresses of the most extreme sort, people are amazed and perplexed.

Perhaps it is not so surprising that the general population undervalues psychosomatic disorders, if you consider for how long they were neglected by doctors and scientists. For most of the twentieth century, neurological psychosomatic disorders, under the names 'hysteria' and 'conversion disorder', were still viewed through a Freudian lens. In *Studies on Hysteria*, his seminal work on the subject, Freud imagined the seizures, paralysis and various disabilities of hysteria as arising from hidden psychological trauma, which is then converted into physical symptoms. For example, a woman too frightened to express herself might suppress the source of her fear, and in so doing lose the power of speech. In Freud's formulation, every symptom could be tracked back to a specific moment of psychological torment. That viewpoint had such traction that, even today, many people, including a large number of doctors, still regard repressed trauma and denied abuse as the full explanation for all psychosomatic disorders. This has created decades-long counterproductive relationships between doctors and patients, in which the doctor insists that their patient is in denial about an unresolved conflict and the patient's subsequent refusal to accept that viewpoint only goes to prove the doctor's point.

The stagnation of scientific progress in the field of psychosomatic disorders left lots of room for a mystery narrative to develop. How is it possible for someone to fall into a coma when their brain seems perfectly healthy? What causes the leg paralysis of a psychosomatic disorder if the nerve pathways are all intact? How does that ethereal thing, referred to as 'the mind', cause seizures? In fact, in the twenty-first century, a considerable amount of work has gone into answering these sorts of questions. In the field of neurology, psychosomatic disorders have become an area of great interest, resulting in a rapid growth in research on the subject. In the scientific world, at least, the one-dimensional concept of stress converted into physical symptoms has been debunked and replaced by more sophisticated explanations. The problem is that these advancements haven't yet made their way very far beyond specialist doctors and patient groups, into the wider public conversation.

What was once called 'hysteria' is now referred to by some as a conversion disorder, or, more recently and more aptly, as a functional neurological disorder (FND). In most medical specialties, the term 'psychosomatic' is still used to indicate medical problems in which the bodily symptoms are thought to have a psychological cause. However, in neurology, the word 'functional' has increasingly replaced 'psychosomatic'. 'Functional' is considered preferable because it indicates that there is a problem with how the nervous system is functioning, while disposing of the prefix 'psych', which is too often (wrongly) distilled to mental fragility, or even madness, in people's interpretation. 'Functional' implies a biological problem, but without the assumption of the presence of stress that has existed in all previous versions of these types of disorders. It leaves open the possibility that emotional trauma is not the only means by which psychological processes can affect the functioning of the brain to lead to disability.

Both psychosomatic disorders in the realms of general medicine and functional neurological disorders in neurology are incredibly common and have the potential to be very serious medical problems. Yet people aren't always aware of that, because, in the public arena, they can be very hard to spot, hidden as they are behind euphemisms, clichés and misunderstanding. That is aptly demonstrated by FND's media portrayal, where it is commonly referred to as a medical 'mystery'.

In 2019, the *Mirror* newspaper ran a story with this headline: *Mysterious illness causes girl to have seizures and wake acting like a toddler*. The article told the story of Alethia, a ten-year-old schoolgirl from Lincolnshire, who had developed limb weakness and seizures. The problem started with pain in her feet and progressed to light and noise sensitivity. In time, her muscles became weak, until, at her worst, she couldn't even lift her head off her pillow. Her symptoms culminated in the development of regular seizures that were followed by odd childlike behaviour. She forgot how to use a knife and fork, and had to be fed like a baby. Despite the 'mystery' headline, Alethia had seen a neurologist and had been given a definitive diagnosis: non-epileptic seizures (one of the many names for psychosomatic seizures) and a functional neurological disorder. Clearly the journalist who wrote the piece did not recognize these diagnoses as legitimate medical conditions, because alongside them he stated, 'medical tests have not been able to provide any answers'.

After reading about Sophie and Alethia, I started wondering why the phrase 'mystery illness' had become such a staple of media reports about psychosomatic and functional medical disorders. It can't be the fact that we do not fully understand the cause and can only make educated guesses at the biological mechanisms that produce psychosomatic symptoms, because there are numerous neurological conditions for which we don't know the cause. Multiple sclerosis, motor neurone disease, Alzheimer's disease – we can't fully explain any of these, we can't cure them, we don't know why they happen, but we don't use nearly as much enigmatic language when referring to them. We call them by their names.

Objective medical investigations are normal in people with functional neurological disorders, even in those with severe disabilities. People paralysed by FND have normal scan results. Brainwave recordings (EEG) are normal in comatose people. The cause of multiple sclerosis may be unexplained, but at least those affected can be reassured that there are demonstrable areas of abnormality in the brain and spinal cord, visible with MRI, to account for and validate their suffering. But the lack of proof in tests can't be what has created the mystery element in the FND storyline either. A migraine doesn't show up on scans, but it is not usually referred to as a mystery illness. FND is a clinical diagnosis, but it is far from alone in that. Until very recently, there were no tests to help with a diagnosis of Parkinson's disease; doctors based the diagnosis entirely on the medical history and clinical examination, and nobody considered this mysterious. They didn't reject the diagnosis just because there were no objective tests to prove it. Are psychosomatic and functional neurological disorders being held to a different standard?

It seems to me that the moniker of mystery is most likely to arise when we are faced with any medical illness that is linked to the concept of 'the mind'. Most people are aware of the connection between emotion and common physical changes like tears and blushing, but cannot extrapolate to more extreme interactions between cognitive processes and physical well-being. We are aware that we can train our brains to allow us to solve mental challenges, like a game of chess, and to master complex physical tasks, like a game of football, but when we try to imagine that the brain can also unlearn those things, it seems a step too far. Yet, if one set of behaviours can teach you a new skill, then surely another set of behaviours could dismantle that skill? That is the fundamental process through which many psychosomatic and functional disorders develop.

There are a great many unanswered questions in the field of psychosomatic medicine, but just as many exist for hundreds of other neurological problems. Yet it is mainly FND, and psychosomatic disorders in general, that still have to fight to shake off centuries-old formulations in order to be seen as legitimate medical conditions.

Reading Sophie's story, it seemed to me very much as if she was a traumatized child who had pulled her physiological shutters down on the world. Loss of awareness without brain disease can be explained by a physiological and psychological process called dissociation – a disconnection between memory, perception and identity which can create a variety of experiences, including feelings of depersonalization and symptoms such as dizziness, blackouts, memory loss and even dissociative (psychosomatic) seizures. But is a pure psychological mechanism, much as Freud might have described, really enough to account for Sophie's coma? Is hers just a physiological stress response, albeit a severe one?

I have seen many people like Sophie, unconscious for hours or even days as a result of dissociation. It affects adults and children. But that is not to say that Sophie's story lacked *all* mystery for me. There were features of what happened to her that I had never seen before. Sophie hadn't moved at all for over a year. She hadn't once opened her eyes. I have never come across a case of dissociation-mediated unconsciousness quite as profound as that. Even more intriguing, or worrying, was that Sophie was not the only one. There were lots of children like her – but only in Sweden. Between 2015 and 2016, 169 children in disparate towns in Sweden had gone to bed and not got up again. Here was a medical problem unique to children and clustering in a single country. If I blamed Sophie's problem on psychological distress belonging to her, arising out of a physiological process inside her head, how would that fit with this strange geographical clustering?

Western medical doctors are trained to interpret symptoms in a very literal way and to treat illness as personal. If somebody has a pain in their chest, we search the heart and lungs for a cause before we consider other possibilities. If we decide that the problem could be psychological in nature, we then look at that person's emotional life for an answer. No doctor is naive to all the external factors that influence illness, but we can't control a person's environment, so we focus our attention on what is within our reach – the person in the room with us. The doctor-patient relationship is an innately intimate one and our system of medicine restrains us within the walls of our institutions and the limits of our specialist qualifications. The news story of Sophie and 168 other comatose Swedish children came as a reminder to me of how much I am neglecting the external factors that have shaped my patients' experiences. I have learned a great deal from listening to the personal stories of people with functional neurological disorders, but maybe I needed to broaden my view.

In 1977, the US psychiatrist George Engel criticized the tendency for doctors to view illness in solely or predominantly biological terms. In a paper published in the medical journal *Science,* he reminded the medical profession that behaviour occurs in a context, and therefore people should never be viewed out of that context. He suggested a new medical model, which he called 'biopsychosocial medicine'.

Every medical problem is a combination of the biological, the psychological and the social. It is only the weighting of each that changes. Cancer is a biological disease, which takes a psychological toll, and which has social triggers and implications. Some cancers have an environmental cause. All of them have an impact on a person's place in the world. All have the potential to wreak biological havoc and to create severe mental distress. Compare cancer with reactive depression, which is triggered by a stressful life event. Reactive depression is a predominantly psychological illness, but it has a biological and social impact too. It can cause weight loss, weight gain, high blood pressure, insomnia, hair loss and many other bodily changes. The low mood is mediated by chemical changes in the brain, but it is caused by and integrally linked to societal factors. It affects the quality of a person's interactions within the world and depends on the response of other people for recovery. Cancer and depression are both biopsychosocial disorders, but have different proportions of each element. Engel encouraged doctors not to forget the social dimension of illness.

Psychosomatic disorders are concerned with the 'psyche', which refers to the mind, and the 'soma', referring to the body. The mind is a function of the brain and is created from biology; it is not the intangible independent entity that Descartes imagined floating away from the body at the point of death. Memory, awareness, perception and consciousness are all integral parts of the mind, and each of these, even if not fully understood, has some measurable neural correlate. But many would say that leaving the environment out of that description of the mind is as foolish as neglecting the impact of society on health. The philosopher David Chalmers described a thought experiment to illustrate that the mind extends into the environment. In it, he told the story of Otto and Inga, two fictional characters who are asked to travel to a museum. Otto has dementia, so he uses directions recorded in a notebook to get there. Inga is healthy, so the directions are stored in her memory. Otto's mind, therefore, has extended into his notebook and the notebook has taken the place of the cognitive process he has lost. Otto's notebook and Inga's memory are serving the same function.

Even though few doctors would disagree with a biopsychosocial concept of illness, and most have moved far beyond limited concepts of the mind, the system of modern medicine doesn't always leave much room for us to incorporate these ideas into our practice. Hospital doctors are so specialized now that many of us only deal with a single organ, never daring to meander outside the field of our expertise. Some doctors are more holistic – general practitioners, in particular – but even they are pushed towards an emphasis on biology. For my part, I would say that I have been neglectful of the external influences on psychosomatic (functional) disorders for the simple reason that they have often felt too big to contemplate. Family and peer-group influences are approachable, but what of all the others – education, religious beliefs, cultural traditions, systems of healthcare and social services, mainstream and social media, the government? When I read about the Swedish children, crystallized into that single extreme example was the reminder I needed of just how much society and culture matter in the shaping of illness, and how much can be learned from observing their effect. The answer to psychosomatic and functional neurological disorders doesn't necessarily sit inside a person's head.

One hundred and sixty-nine children, in a single small geographical area, fell into a coma for what is believed to be a psychological reason. That means that 169 brains have each been pulled or pushed or moulded into behaving in a unique and unusual way. With all of the victims so geographically contained, there simply must be something in their social environment that has created that possibility.

In 2018, I went to Sweden to visit children like Sophie and realized that mass illness outbreaks like theirs, ones that happen in small communities, have a great deal to say about how societal and cultural factors affect biology and psychology to create psychosomatic and functional disabilities. They put a magnifying glass to the social elements that affect health. That first journey would lead me to a number of other equally intriguing cases, including a Nicaraguan community in Texas who pass seizures from generation to generation through an inherited narrative; to a small town in Kazakhstan where over a hundred people 'inexplicably' fell asleep for days on end; to Colombia, where hundreds of young women's lives have been disrupted by seizures; and to upstate New York, where a media frenzy had a drastic effect on the health of sixteen high-school students. As I travelled the world and trawled through newspapers, I found recurring themes in the accounts of very different populations, and even in the strangest stories I heard a great deal that reminded me of my own patients. The Swedish children are far from unique in being caught up in a health crisis that is specific to a time and place. Outbreaks of mass psychosomatic illness happen all over the world, multiple times per year, but they affect such unrelated communities that no one group gets the chance to learn from another.

Sophie is part of a group, and whatever that group shares must have been pivotal in creating their shared coma. But Alethia is part of a group too, even if she doesn't know it. There are hundreds of thousands of people like her, some of whom are lucky enough to have a diagnosis, but many are resigned to living with the label 'mystery illness', without realizing that medical science has moved on, there are explanations available and, more importantly, there is help out there, if they could only find it.

1

The Sleeping Beauties

Reductionism: *The belief that human behaviour can be explained by breaking it down into smaller component parts.*

I had barely stepped foot over the threshold and I already felt claustrophobic. I wanted to turn back. People shuffled into the room in front of me, while somebody else stood directly behind me, a little too close. It felt hard to escape.

I could see Nola lying in a bed to my right. She was about ten years old, I guessed. This was her bedroom. I had come knowing what to expect, but somehow I still wasn't prepared. Five people and one dog had just walked into the room, but she didn't have so much as a flicker of acknowledgement for any of us. She just lay perfectly still, her eyes closed, apparently peaceful.

'She's been like this for over a year and a half,' Dr Olssen said, as she bent to stroke Nola gently on the cheek.

I was in Horndal, Sweden – a small municipality a hundred miles north of Stockholm. Dr Olssen was my guide. She was a slim, deeply tanned woman in her sixties, with a distinctive triangular white patch in the fringe of her light brown hair. She had been caring for Nola since the child had first fallen ill, so she knew the family well. Dr Olssen's husband, Sam, and their dog had also come with us. All three were regular visitors to Nola's home and knew their way around it. From the front door, they had led me promptly and directly to Nola's room. It was almost too abrupt for me. One moment I was outside in the midday sunshine, then suddenly I was in the twilight of a sleeping child's room. I had an impulse to open the curtains. Dr Olssen must have felt the same, because she walked to the window, drew the curtains aside and let the light in. She turned to Nola's parents and said, 'The girls have to know it's daytime. They need sun on their skin.'

'They know it's day,' her mother answered defensively. 'We sit them outside in the morning. They're in bed because you're visiting.'

This wasn't just Nola's room. Her sister, Helan, who was roughly a year older, lay quietly on the bottom of a set of bunk beds to my left. From where I stood, I could only see the soles of her feet. The upper bunk – their brother's bed – was empty. He was healthy; I had seen him peeping out from around a corner as I walked to the girls' room.

Dr Olssen turned and called to me: 'Suzanne, where are you? Aren't you coming to say hello? Isn't this why you're here?'

She was hunkered down by Nola's bed, brushing the child's black hair to one side with her fingers. I stood wavering near the threshold, struggling to take the final few steps of a long journey. I was pretty sure I was going to cry, and I didn't want the others to see. I wasn't ashamed; I am human and upsetting things upset me. Sick children in particular upset me. But this family had been through so much and I didn't want to put them in the position of having to comfort me. I fixed a smile on my face and approached Nola's bed. As I did, I glanced over my shoulder at Helan, and was surprised to see her eyes open for a second to look at me and then close again.

'She's awake,' I said to Dr Olssen.

'Yes, Helan's only in the early stages.'

Nola showed no sign of being awake, lying on top of her bed covers, laid out in preparation for me. She was wearing a pink dress and black and white harlequin tights. Her hair was thick and glossy, but her skin was pale. Her lips were an insipid pink, almost colourless. Her hands were folded across her stomach. She looked serene, like the princess who had eaten the poisoned apple. The only certain sign of illness was a nasogastric feeding tube threaded through her nose, secured to her cheek with tape. The only sign of life, the gentle up and down of her chest.

I crouched beside her bed and introduced myself. I knew that, even if she could hear me, she probably couldn't understand. She knew very little English, and I didn't speak Swedish or her native language, Kurdish, but I hoped the tone of my voice would reassure her. As I spoke, I looked back at Helan again. Her eyes were fully open and she met my gaze, allowing me to see that she was watching me. I smiled at her, but her expression didn't change. The girls' mother stood at the end of Nola's bed, leaning a shoulder against the wall. She was a striking woman, with high cheekbones and a prominent café-au-lait birthmark on her forehead. She had relinquished control to Dr Olssen and was watching me closely. She seemed calm and dignified. Her husband, the children's father, stood at the doorway, shuffling in and out of the room.

Like Sophie, whose story I had read in that newspaper article, Nola and Helan are two of the hundreds of sleeping children who have appeared sporadically in Sweden over a span of twenty years. The first official medical reports of the epidemic appeared in the early 2000s. Typically, the sleeping sickness had an insidious onset. Children initially became anxious and depressed. Their behaviour changed: they stopped playing with other children and, over time, stopped playing altogether. They slowly withdrew into themselves, and soon they couldn't go to school. They spoke less and less, until they didn't speak at all. Eventually, they took to bed. If they entered the deepest stage, they could no longer eat or open their eyes. They became completely immobile, showing no response to encouragement from family or friends, and no longer acknowledging pain or hunger or discomfort. They ceased having any active participation in the world.

The first children affected were admitted to hospital. They underwent extensive medical investigations, including CAT scans, blood tests, EEGs (electroencephalograms, or brainwave recordings) and lumbar punctures to look at spinal fluid. The results invariably came back as normal, with the brainwave recordings contradicting the children's apparent unconscious state. Even when the children appeared to be deeply unresponsive, their brainwaves showed the cycles of waking and sleep that one would expect in a healthy person. Some of the most severely affected children spent time under close observation in intensive-care units, yet still nobody could wake them. Because no disease was found, the help doctors and nurses could offer was limited. They fed the children through feeding tubes, while physiotherapists kept their joints mobile and their lungs clear and nurses made sure they didn't develop pressure sores through inactivity. Ultimately, being in hospital didn't make much difference, so many children were sent home to be cared for by their parents. The children's ages ranged from seven to nineteen. The lucky ones were sick for a few months, but many didn't wake for years. Some still haven't woken.

When this started happening, it was unprecedented. Nobody knew what to call it. Was it a coma? That word wasn't quite right; it implied deep unconsciousness, but some of the children seemed to have awareness of their surroundings. Tests showed that their brains responded to external stimuli. Sleep certainly wasn't the right word either. Sleep is natural, but what was happening to the children was not – it was impenetrable. In the end, Swedish doctors settled for 'apathy'. The Swiss psychiatrist Karl Jaspers described apathy as an absence of feeling with no incentive to act. It is a total indifference to pain and to pleasure, a complete freedom from emotion of any kind. That description fitted with what the doctors were seeing. After a few years, apathy was converted to an official medical designation – *Uppgivenhetssyndrom* – literally meaning 'to give up'. In English, this became 'resignation syndrome'.

Standing at Nola's bedside, the label seemed apt. Dr Olssen rolled up Nola's dress, exposing her bare stomach and revealing that she was wearing a nappy under her tights. Nola didn't resist the intrusion. Her hand lolled over the side of the bed, the dog nudged it with his nose, but she didn't respond to that either. Dr Olssen pressed on her stomach and listened to it with a stethoscope, and then listened to her heart and lungs. The examination, Dr Olssen's friendly chatter, the stranger in the room, the pacing of the dog – none of these elicited any sort of reaction.

Periodically, Dr Olssen turned to me to report on her findings.

'Her heart rate is ninety-two. High.'

When she said that, I again felt uncomfortable to the point of upset. Ninety-two seemed high to me, too. Ninety-two did not sound like the heart rate that goes with an absence of emotion, in a child who had not moved for over a year. It suggested a state of emotional arousal – in other words, the very opposite of apathy. The autonomic nervous system has unconscious control over the heart rate. The parasympathetic nerves slow everything down when a person is at rest, while the sympathetic system powers our fight-or-flight mechanism, speeding up the heart in preparation for action. What was Nola's body preparing for?

Dr Olssen rolled up Nola's sleeve and tested her blood pressure. The child didn't flinch. 'One hundred over seventy-one,' Dr Olssen told me, which is normal for a relaxed child. She lifted Nola's arm to show me how limp it was. The arm dropped unceremoniously onto the bed when let go. I had read reports that described how Dr Olssen put ice packs on the children's bare skin to see if they showed a response, and I had seen a picture in a newspaper of a child with resignation syndrome who had a packet of frozen vegetables pressed against her exposed stomach. When I was at medical school, we were taught to give painful stimuli as part of the assessment of an unconscious patient. I don't do it anymore; I gradually realized it was an unnecessary cruelty, so I was glad when Dr Olssen didn't suggest testing Nola in this way for my benefit. Instead, she turned to me and asked me to carry out an examination.

I hesitated. I am a doctor, but I wasn't Nola's doctor. I looked at her mother, who was still standing at the end of her bed. We had no shared language. What brief conversation we had went through Dr Olssen. She seemed happy for me to be there, but I longed to talk to her directly, without a go-between. There were so many languages, and such varied dynamics between the people around the bed, that I found it hard to read the room.

Dr Olssen raised her eyebrows as she waited for me to answer. 'What did you come here to do?'

Good question. Suddenly, I didn't know why I was there. I saw patients a little like those girls all the time in my job. What made them special enough for me to feel the need to visit, and what did I hope to gain?

Dr Olssen placed a gentle thumb on Nola's eyelid and drew it open. Nola's eye rolled up, so that only the white was visible. 'Bell's phenomenon,' Dr Olssen said.

Bell's phenomenon is a normal eye reflex that occurs when a person closes their eye: the eye rolls up and outwards as the lid lowers. But Dr Olssen wasn't closing Nola's eye, she was opening it. What I saw was evidence of a child resisting eye opening. Her eye rolled up because she was fighting to keep them closed. Was that an unconscious reflex, or was Nola's resistance to communication more active than passive, after all?

'Come on.' Dr Olssen coaxed me forward. 'You're a neurologist, aren't you?'

I remembered why I was there. Dr Olssen was a retired ear, nose and throat doctor, desperate to help the children and support the families. She'd welcomed me because I was a neurologist. She hoped that I could find an explanation for what had so far been inexplicable; that I would interpret the clinical signs and, by doing so, legitimize the girls' suffering and convince someone to help them. That Nola had been lying in bed for a year and half without eating or moving had not been deemed impressive enough to get her the help she needed. A neurologist, a specialist in brain disease, would add weight to the diagnosis, or so Dr Olssen hoped.

That's how modern medicine works: disease impresses people; illness with no evidence of disease does not. Psychological illness, psychosomatic and functional symptoms are the least respected of medical problems.

'Examine her,' Dr Olssen said again.

Reluctantly, I took Nola's legs in my hands and felt the muscle bulk. I moved her limbs to assess mobility and tone. Her muscles felt healthy, not wasted. Her reflexes were normal. Apart from her unresponsiveness, there was nothing abnormal.

I tried to open Nola's eyes, as Dr Olssen had, and felt her resist. Dr Olssen asked me to palpate the muscles in her cheeks. In contrast to every other muscle in her small body, these were rigid. Her teeth were clenched shut – another piece of evidence against passive, apathetic restfulness.

I looked behind me, at Helan. The dog was staring at her; Sam, Dr Olssen's husband, was holding him by the collar to keep him in check. Helan looked past the dog, at me. I smiled at her again, but she just stared back blankly.

Dr Olssen followed my gaze. 'Nola was the first to get sick. Helan only got symptoms after the third asylum refusal, when the family were told they had to leave Sweden.'

Despite Dr Olssen's interest in uncovering the brain mechanism to explain the children's apathy, everybody – family, doctors, officials – knew why Nola and Helan were sick. And they knew exactly what was required to make them better.

Resignation syndrome is not indiscriminate. It is a disorder that exclusively affects children of asylum-seeking families. These children were traumatized long before they fell ill. Some were already showing very early signs of illness when they arrived in Sweden, but most only began to withdraw when their families were faced with the long process of asylum application.

Nola had come to Sweden when she was two and half – at least, that was the official age she was given on arrival, by a man she had never met before. Nola's family fled the Turkey–Syria border when she was a toddler, and their journey to Sweden had been uncharted. Somewhere in transit, their papers were destroyed. Arriving at the Swedish border, they had no proof of who they were or where they came from, so the authorities estimated their ages. They determined Nola to be two and a half, Helan to be three and a half, and their younger brother to be one.

Nola's family are Yazidi, an ethnic-minority people indigenous to Iraq, Syria and Turkey. The worldwide number of Yazidi is estimated to be fewer than 700,000. Walking through the house to Nola's room, I had seen a picture of a peacock hanging on the wall, dark blue with his open tail feathers displayed behind him. Nola's father had a peacock tattoo on his arm. The Peacock Angel is central to the Yazidi religion. They believe he was created by the supreme deity and that he governs earth. The stories told about the Peacock Angel have links to the beliefs of other religions. He is said to have taught Adam and Eve. He is also the reason that the Yazidi have been referred to as devil worshippers. Some say that, because the Peacock Angel rebelled against God and was cast into hell, he therefore represents Satan. It is this sort of interpretation of their beliefs that has seen the group subjected to centuries of persecution. In the nineteenth and twentieth centuries alone they were subjected to seventy-two genocidal massacres, while in the twenty-first century they have been the victims of many bloody attacks, first in Iraq and more recently in Syria. Women and children have been gang-raped and taken as sex slaves. In the region of 70,000 Yazidi people are said to have sought asylum in Europe.

Nobody can prove what Nola and her family suffered before they came to Sweden – I can only tell you the story I was told. The family used to live in an underdeveloped rural village in Syria, near the border with Turkey. Most of the people had no running water, but they had a communal well to which Nola's mother made daily trips. One morning, when she went to get water, she was grabbed by a group of four men, who dragged her into the woods and assaulted her. When she came home and told her family what had happened, her father was furious with her for bringing shame on them. Over the next weeks, there were heated arguments between Nola's grandfather and her parents. In one of them, Nola and her siblings were in the room when their grandfather threatened to kill their mother. At the time of the assault, Nola's mother had been pregnant with her fourth child, but she soon miscarried.

With threats to the family from inside and outside the home, staying in Syria was untenable and the family were forced to flee. Arriving in Sweden with no papers, unable to speak Swedish and unable to read a Latin-script alphabet, they struggled to communicate and had no way of verifying where they'd come from or who they were. They immediately applied for asylum, but asylum depended on them proving they had been persecuted in their country of origin and convincing the authorities that it was unsafe for them to return.

At the time, Sweden took a generous stance on asylum seekers, and Nola's family were given temporary permission to stay. The subsequent process of applying for permanent asylum was very slow. Before it was properly underway, Nola and Helan were already in school. After several years, the family's application for asylum was processed – and then refused, although they had the right to appeal the decision, not once but twice. By that time, the Syrian war had started, making their birthplace even more dangerous. It was at this point that Nola showed the first signs of withdrawal.

The children had lived in Sweden for longer than they had lived anywhere else. All their friends were here, both children spoke fluent Swedish and Helan also had a good understanding of English. I don't know what Nola and Helan knew of the place they were born, but, even if it was never explicitly discussed, they must have felt the fear associated with returning there. The family had placed themselves in great danger fleeing Syria, and – whether they were believed or not – they had done so for a reason.

'I will ask the father to show you what Nola is like when they try to get her out of bed,' Dr Olssen said. On her instructions, Nola's father sat his daughter up and swung her legs over the side of the bed. Her body was like a rag doll. Her head hung down onto her chest. Her father stood behind her, holding her under the armpits so that her shoulders hunched and her hands dangled by her sides; then, encouraged by Dr Olssen, he walked her like a marionette across the room. Nola's feet dragged behind her, toes scuffing the carpet. I knew why I was being treated to this almost grotesque display. We often require a person to look sick and, ideally, to have at least one abnormal objective medical test to prove that they are genuinely ill. Nola's tests were all normal, but Dr Olssen and Nola's family wanted me to understand how bad things were for her – so they showed me.

'The children are the ones who open the letters,' Dr Olssen said, as Nola's father laid his daughter back on the bed and her mother made her comfortable.

'Pardon?'

'The parents can't speak Swedish, so when letters arrive from the immigration office it is usually the children who read them. The children translate for the parents.'

'That seems so wrong.'

'They are their parents' conduit to the new world.'

'There must be a better way.'

Dr Olssen laughed. 'You are very naive. You can't stop the children from noticing.'

'I suppose not.' I was thinking of the children in my family, innocent and sheltered. I looked at Helan. She was so young, yet had been through so much. 'How old is Helan?' I asked.

'Eleven, but they contested that, too.' Dr Olssen grimaced. 'The school said that she spoke in an adult way and couldn't possibly be the age that the family claimed.'

Correctly applying ages to people seeking asylum has caused problems worldwide. Children are mistakenly placed in facilities with adults if they appear older. Claims are made that adults pretend to be children so they will be treated more leniently. But the medical assessment of age is fraught with error. There is no reliable way of knowing how varying degrees of chronic deprivation, abuse, malnutrition, not to mention the process of fleeing and of seeking asylum might affect a person's physical appearance, bone age, muscle bulk, sexual maturation, behaviour or language.

Helan was unequivocally a child, even if her official age was up for debate. She was prepubertal. Like both Nola and her mother, she had a long mane of thick black hair. Yazidi women do not cut their hair. As I watched her, I saw her eyelids fluttering closed and then opening again repeatedly. I had been told she spoke English, so I kneeled by her bed and told her my name. To my surprise, she whispered something back. It was very quiet, so I asked her to repeat it and I leaned in close so I could hear. She said her name: 'Helan.'

Helan had only been sick for a few months, since the family's third and final application for asylum had been rejected.

'When the third rejection letter arrived, Helan said, "What will happen to my sister?"' Dr Olssen told me. 'Then she became quiet and we saw she was getting sick. I told her parents not to let her stay in bed. I told them to make her eat and to keep her in school, but it was impossible.'

The authorities believed that the family was Turkish, Dr Olssen told me. They could not be sent back to war-torn Syria, but if they were Turkish they could be returned there. I couldn't imagine how it must have felt to a child to be told they would have to leave their home to go to a place that existed only as a horrific story. Driving up to the family's apartment block, along a wide tree-lined road, I had been struck by what a lovely area it was. The three children shared a room, but otherwise their apartment was spacious and it overlooked a leafy playground. The children's room had drawings on the wall and an assortment of books and board games piled in one corner. The games looked well played with – although not by these children, I thought.

'Her friends from school still visit her. One girl comes and reads to her every week,' Dr Olssen told me, before turning to Helan. 'Would you like a story now?'

The little girl nodded.

Dr Olssen took a picture book from the pile and started to read. The girls' brother peeped around the door shyly. At first, I thought he was listening to the story, but then I realized it was the dog that was holding his attention. Sam saw it too and asked if he would like to take the dog for a walk, and the two disappeared together happily.

Sam knew the families of the resignation-syndrome children as well as his wife did. He was a kindly white-bearded grandfather, a US citizen by birth and a trained psychologist, although in Sweden he worked in IT. Between them, the couple provided support to fourteen children from seven families, each child in a different stage of resignation syndrome. The previous evening, I had stayed at their home, a lived-in wooden house full of books, plants and family pictures. While eating scones and drinking peppermint tea in their garden, Sam had told me how they'd seen that, when granted residency, the children usually woke up, albeit not overnight. The road to getting better was as gradual and tedious as the onset of apathy. It could take months or longer, depending on how long the child had been ill.

Although there were no miraculous awakenings, once recovered, the children were able to flourish in their new lives. Another child they cared for, a girl called Aliya, had fled from a former Soviet republic as a member of a persecuted minority. She'd suffered from resignation syndrome for over a year, showing no signs of voluntary activity until the day after she was told she had permanent residency in Sweden, when she opened her eyes for a moment. Over the weeks that followed, she awoke fully and, after a few months, she was back at school. Despite entering the Swedish school system late and having large gaps in her education, she excelled in her exams and was now studying for a law degree.

'Did she tell you what it felt like to suffer from resignation syndrome?' I asked.

I had been wondering about this since I first heard about the condition. The newspaper articles had made the disorder sound completely passive, but I suspected that the words *apathy* and *resignation* did not really capture the experience of those suffering from the syndrome at all.

'She doesn't like to talk about it,' Sam told me. 'That's the same for all the children. Once they come out of it, they just want to put it behind them.'

I still really needed to know, so I asked again: 'Did she tell you if she was aware of what was happening, or was it more like losing time? As if she had just been asleep?'

'She said it was like being in a dream that she didn't want to wake up from.'

I liked that description. I understood it and it made the children's plight less frightening. I had read of one young boy's experience in a magazine, and it had been much more disturbing: 'he had felt as if he were in a glass box with fragile walls, deep in the ocean. If he spoke or moved, he thought, it would create a vibration, which would cause the glass to shatter. "The water would pour in and kill me," he said.'

Aliya often stayed with Dr Olssen and Sam, in their home.

Even when the children recovered, this generous couple never stopped supporting them and their families. They found jobs for those in need of work and helped them with applications for school and college. They even allowed whole families to stay with them if they had nowhere else.

I watched Dr Olssen, perched on the side of Helan's bed, reading a story in a language I didn't understand. Every now and then, she turned the pages of the book towards Helan, so the child could see the pictures. Meanwhile, the girls' mother sat on the edge of Nola's bed, using a hairbrush to gently brush the child's bare arms and legs. This was their daily routine, a sensory experience. Afterwards, she moved Nola's joints, bending and straightening her knees and elbows, rolling her hips and shoulders and wrists in turn. These were exercises that she had learned from a physiotherapist and which would stop the children developing contractures – stiff, shortened tendons, caused by immobility.

I looked from girl to girl. Nola was pale, but other than that she looked in perfect health. Helan had colour in her lips and skin, reflecting how much more she moved and the shorter term of her illness. The family washed and dressed the girls every day. They tried to create a routine, so the girls would have a sense of morning and afternoon and evening. They changed the girls' position in bed to protect their skin from ulcers. They sat them in wheelchairs by the dinner table, so they wouldn't forget they were part of a family, and they placed titbits of food on their tongues in the hope of tempting them to eat. They wet their lips with water. They tried giving each girl a straw to drink from – which Helan could manage, but Nola ignored.

The care routine and bedside story created a gentle, loving scene. Occasionally, Helan mouthed something as Dr Olssen was reading, and I realized she was reciting parts of the story that she knew off by heart. When the story was finished, Dr Olssen promised Helan a new story at the next visit. As we left their room, both girls lay just as we had found them.

The earliest official reports of resignation syndrome appeared in 2005. Rumour suggested it had been around since the 1990s, but the number of children affected escalated at the turn of the century. Between 2003 and 2005, 424 cases were reported. There have been hundreds more since. It affects both boys and girls, but with a slight preponderance of girls.

The first children to fall ill were usually admitted to hospital, where they underwent medical investigations and treatment. Once test results came back as normal, there were the inevitable accusations that the children were pretending to be sick, which is so often the fate of people whose physical disability cannot be explained by an organic disease process with measurable biochemical or structural anatomical abnormalities. But, flying in the face of that suggestion, children as young as seven remained completely unresponsive, even during long-term hospital admissions. Many of the children have been subjected to medical testing and inpatient care under the supervision of a variety of specialists. Some early cases were admitted to intensive-care units, separated from their parents and kept under close medical scrutiny, but they still didn't wake up. No child could sustain such a prolonged apathetic state voluntarily.

Some people's attention turned to the parents. Were the children being sedated? Or even poisoned? One media report suggested that a doctor had seen a parent giving a child liquid medicine. But that is easy to check, and blood and urine samples showed no evidence of any intoxicant. Some still said that this was Munchhausen's syndrome by proxy – a type of child abuse in which a parent or carer fabricates illness and seeks unnecessary healthcare for their child. Supporters of that theory postulated that parents were coaxing or coercing the children into developing the condition. One doctor said that the families were using the children as Trojan Horses to gain admittance to a new country for the rest of the family. There were claims that nurses had seen children who were supposed to be unresponsive fighting against attempts to insert a nasogastric tube.

Could Nola be roaming around the apartment when the curtains were pulled? Did the two little girls jump into bed, under instruction from their parents, whenever a visitor knocked on the door? When I visited Sweden in 2018, other than suspicious rumours in newspaper articles, there was not a grain of evidence to support this view. Then, in October 2019, an adult came forward and said that, as a child, she had been coerced into being 'apathetic' by her parents. That created a brief furore of accusations. It risked making every family look guilty. But there will always be people who cheat disability payments and insurance companies, people who take advantage of situations for their own gain; that should not be seen to imply that every member of a group is guilty. Other than that sad case, which amounts to child abuse, no official enquiry has recorded deceptive behaviour in the resignation-syndrome children or their parents. Even children admitted for long-term care into psychiatric units displayed no behaviour that would support the accusation of Munchausen's syndrome by proxy. Children did recover once they had been offered asylum and hope was restored to their lives, but that recovery was in line with recovery from any chronic serious illness – gradual and dependent on the length of illness and degree of disability present.

People who have psychologically mediated physical symptoms always fear being accused of feigning illness. I knew that one of the reasons Dr Olssen was desperate for me to provide a brain-related explanation for the children's condition was to help them escape such an accusation. She also knew that a brain disorder had a better chance of being respected than a psychological disorder. To refer to resignation syndrome as stress induced would lessen the seriousness of the children's condition in people's minds. It is the way of the world that the length of time a person spends as sick, immobile and unresponsive is less impressive if it doesn't come with a corresponding change on a brain scan.

Biological correlates are often used to give credence to the experience of psychosomatic disorders. An objective change on a blood test or scan allows others to believe in the suffering. It is not surprising, therefore, that a great deal of thought has gone into trying to understand the biomechanics of resignation syndrome. Not only would it be of scientific interest and guide treatment, but, more than anything else, understanding the biomechanics would validate the children's level of disability. Various incomplete theories have attempted to shed light on the biology of the disorder. Doctors have noted a fast heart rate and high body temperature in some of the children, which seems to suggest that a stress response mediated by hormones or the autonomic nervous system could play a part in the disorder. A single small study looking at four children showed a lack of the normal daily variation in the level of the stress hormone cortisol, lending a little weight to the stress hypothesis. In the same vein, one group of scientists speculated that stress hormones in pregnancy affected brain development and reduced the children's ability to cope with stress later in life. The problem with these observations and theories is that neither stress hormones, the autonomic nervous system, nor poor brain development would account for the unusually sustained and profound physical manifestations of the disorder, nor the strange geographical distribution. There are asylum-seeking families all over the world, but none have responded to their situation like the children in Sweden. Stress is common, but resignation syndrome is not.

Some scientists likened the disorder to catatonia, a condition in which the affected person is immobile, with little or no physical response, but remains aware. Catatonia can be caused by brain disease, but it also occurs in the context of psychiatric conditions. Although poorly understood, it has been associated with a variety of neurotransmitter and brain-scan irregularities. Specialist brain scans have shown metabolic changes in regions of the frontal lobes in people in a catatonic state, and Swedish scientists are keen to carry out more detailed brain scanning on the resignation-syndrome children to find out if this is also the case for them. While there is something in the description of catatonia that resonates with what is happening to Nola and Helan, resignation syndrome lacks the characteristic stiffness and posturing of that condition. Catatonic patients look like animals posed by taxidermy; Nola was a rag doll.

One ongoing uncertainty has been whether or not the children are aware of their surroundings, particularly as they seem unwilling or unable to clear up this mystery themselves. They are sometimes referred to as being in a coma, but many have been observed to cry or exhibit occasional eye contact, which suggests otherwise. Patients in persistent vegetative states have been assessed for consciousness using functional MRI, so scientists have proposed testing in that way to solve the 'awareness' dilemma in resignation syndrome.

Not all the medical interest in this disorder has focused on blood tests and brain scans. More psychologically minded explanations have compared resignation syndrome to pervasive refusal syndrome (also called pervasive arousal withdrawal syndrome – PAWS), a psychiatric disorder of children and teens in which they resolutely refuse to eat, talk, walk or engage with their surroundings. The cause is unknown, but PAWS has been linked to stress and trauma. The withdrawal in PAWS is an active one, as the word 'refusal' suggests; it is not apathetic. Still, as a condition associated with hopelessness, it does seem to have more in common with resignation syndrome than other suggestions.

The resignation-syndrome children became ill while living in Sweden, but most had experienced trauma in their country of birth. It seems likely, then, that this past trauma would play a significant role in the illness. Perhaps it is a form of post-traumatic stress disorder? Or could the ordeals suffered by the parents have affected their ability to parent, which in turn impacted on the emotional development of the child? One psychodynamically minded theory is that the traumatized mothers are projecting their fatalistic anguish onto their children, in what one doctor described as an act of 'lethal mothering'.

There is clearly much of value in investigating both the psychological and biological explanations for resignation syndrome, but even when taken together they fall short. Psychological explanations focus too much on the stressor and on the mental state of the individual affected, without adequately paying attention to the bigger picture. They also come with the inevitable need to apportion blame, passing judgement on the child and the child's family. They risk diminishing the family's plight in the eyes of others. Psychological distress simply doesn't elicit the same urgent need for help that physical suffering does.

But the biomedical theories are even more problematic. The search for a biological mechanism is in part an attempt to ensure that the children's condition is taken seriously, but it also threatens to neglect all the external factors that have propelled the children into chronic disability. MRI scans that try to unpick the brain mechanism of resignation syndrome are useful research tools and might offer general insights into how the brain controls consciousness and motivation, but there is something faintly ludicrous about expecting scans done on individuals to explain or solve a group phenomenon.

As a neurologist, people expect me to be especially interested in the brain mechanisms that cause disability. But, standing in the bedroom shared by Nola and Helan, the confused neural networks keeping these small children in bed seemed only to be an end point and, therefore, the least important part of what created their situation. A whole lifetime had led Nola and Helan to this place, where they lay in the confines of a Swedish bedroom, the curtains pulled on a sunny day.

Biological and psychological hypotheses for resignation syndrome are reductionist, in exactly the way that Engel's biopsychosocial theory tried to address. They focus on the inside, while failing to incorporate what is on the outside – the odd geographical clustering. In fact, there is even more to the story that helps demonstrate the futility of an overly individualistic approach to the children's situation. Not only is resignation syndrome restricted to children seeking asylum in Sweden, it is restricted even within that very specific group. It doesn't affect all asylum seekers; children from countries of the former Soviet republics and from the Balkans are more likely to suffer from it. Yazidi and Uyghur people, ethnic groups that have recently been subject to a great deal of persecution, are also disproportionately affected. It has not been reported in refugees who originated in Africa, and rarely in any other nationality or ethnic group.

If psychogenic and biomedical theories fully accounted for

the cause of resignation syndrome, then why do we not see it happening all over the world? And why doesn't it happen to people of different ages and backgrounds? Psychological trauma and hardship exist in every society, and all our brains are biologically the same. That the disorder picks off its victims so selectively shows the error of viewing it as only a biological problem, concerned with hormones and neurotransmitters, or a psychological problem, linked to the personality of an individual.

It seemed obvious to me, having heard the girls' story, that there was something to learn from the cultural specificity of the disorder. It suggested that resignation syndrome may not be a biological or psychological illness, in the Western sense; it may in fact be a sociocultural phenomenon. If so, then brain scans and cortisol levels would be largely meaningless.

After our visit to see Nola and Helan, Dr Olssen and Sam took me back to their home to stay for the night. Sitting on the terrace, eating a dinner of salmon and salad, I could see red wooden cottages and barns punctuating a rolling green landscape. As we ate, we talked about the children, and I took the opportunity to raise the issue of the strange clustering of the disorder in this small group of marginalized people. Dr Olssen did not look happy at that. I sensed I was a disappointment to her. I was not the great neurologist who would provide the perfect biological explanation for resignation syndrome and then write to the Swedish Migration Agency and secure asylum for every child who needed it.

'It's not happening because they are Yazidi,' she said, when I touched on the subject.

But that was not the culture to which I was referring. Her own culture, now shared with the children, interested me just as much as their country of origin and ethnicity. Asylum seekers of Yazidi or Uyghur ethnicity, and those of Balkan or Soviet origin, do not get resignation syndrome when they flee to countries other than Sweden. If societal influences lead to this disorder, they do not stem from the country of origin, but rather from some combination of circumstances. The vulnerabilities created by the children's past experiences were important, surely, but so too was their journey to and their life in Sweden. After all, Nola and Helan had spent the vast majority of their lives there.

Sweden had been welcoming to the family when they arrived. They were granted temporary residency and given a home pending their application for asylum. The process took three years to get started in earnest, at which stage both girls were in school. They spoke fluent Swedish. They had established friendships. I wondered if they knew that people saw them as different and that their home was potentially only temporary. Once underway, the application process was drawn out over several years. Although the family was not on trial, they felt as if they were interrogated rather than listened to. The asylum system seeks to find the mistakes that disprove an applicant's case, rather than looking for evidence to prove it.

Asylum-seeking processes all over the world are subject to similar problems. Families making an application have usually come from a place with a poor human-rights record, where the authorities cannot be trusted. The application requires asylum-seeking families to face panels asking them to prove that their stories are true. Questioning can be intrusive and combative, and rarely takes into account what the applicant has already suffered. They are forced to defend their stories in an intimidating environment, after which a small group of people decide whether those stories are credible or not credible. The children are usually obliged to be present for the hearings.

Sweden has a reputation for being liberal, less racist and less hostile to immigrants than many other countries, and, until fairly recently, families with children suffering from resignation syndrome were offered automatic asylum. In 2014, the Swedish prime minister asked the population to 'open their hearts' to asylum seekers, and there followed a year in which a record number of foreign nationals arrived in the country. But the mood soon changed: in keeping with a worldwide trend, Sweden saw a rise in right-wing politics and anti-immigrant rhetoric, and the number of new asylum cases subsequently dropped drastically. It's possible that hostility to immigrants and heightened tensions have contributed to the spread of resignation syndrome, but there were decades of apathetic children before the tide turned against asylum seekers. Given that resignation syndrome is considered by many to be caused by hopelessness, and can therefore be treated by the restoration of hope, it is perhaps not far-fetched to conclude that the lengthy asylum process, through three stages of hearings, could be a contributing factor to the development of the disorder. Nola and Helan have spent almost their entire lifetimes alternating between anticipation and despondency. That has physical consequences.

The internal mechanism for coping with stress is the same for the children as it is for other people; it is only their lives that are different. Surely it makes sense, then, that it is something in their environment that has shaped their physical reaction to psychological distress in such a unique and specific manner. The disability must arise in the brain – because all behaviour does – but it is the manner in which external factors impact on the children's brains, not innate physiology, that matters most to the development of resignation syndrome.

A child faced with deportation is a child under strain. That creates physical symptoms by stimulating physiological stress pathways. Up to that point, the asylum-seeking child's physical experience is the same as that of any person experiencing stress or trauma. What happens next is what makes the outcome for children like Nola and Helan different from that of everybody else, and this stage is heavily dependent on life experience.

The chain of events that has created resignation syndrome cannot belong to a single child or a single family. The key to understanding it is unlikely to lie inside the head of a child. It is far bigger than that. The children are vulnerable people, made so by a history of deprivation and stress. Their response to the asylum experience is unlikely to be purely biological, but rather comes from an expectation that has been programmed in them, with contributions from all the people in their environment – in their country of origin, but even more so in Europe. The children are embodying a sociocultural phenomenon. Their story has been written across nations, in a combination that has made them unique. It has been impacted by poor social circumstances, poor nutrition, epigenetics, abusers, authority figures, politicians, parents, doctors and the media. Without the correct combination of these, resignation syndrome would not exist.

There are more than two sides to all psychosomatic and functional illnesses, not just resignation syndrome. The cultural specificity of resignation syndrome acted as a crucial reminder to me that the social element should never be underestimated in other similar disorders. In day-to-day medicine, most doctors are very aware of the role of society in illness, but it is not always easy to talk about or address. It is a huge and intimidating problem, outside of the scope of individual doctors, but that means that something is being missed. Psychological and biological factors are fundamental in psychosomatic and functional disorders because they produce the symptoms, but sociocultural factors can be even more important. They may direct the course of illness, shape the symptoms and determine the severity and the outcome; they tell a person where to go for help and even dictate treatment. The resignation-syndrome children are told they will not wake up until they are granted asylum, and so it unfolds. They are *unconsciously* playing out a sick role that has entered the folklore of their small community.

The children are a small group, but there is much to be learned from their situation. For too long, the role of social environment in the development of illness has been neglected in favour of psychological and biological perspectives. However, I am a pragmatist, and as such I am aware that that sort of insight won't necessarily feel helpful to people with these types of disorders. Dr Olssen worried that portraying resignation syndrome as a sociocultural disorder felt more like judgement than enlightenment. If I had been able to talk directly to the parents without an interpreter, I think they would still have wanted to know the biology, because ultimately that's what everybody wants to know.

The thing that stands to be understood is exactly how external factors were able to change biology to create the very specific clinical features of resignation syndrome. Meeting Nola and Helan set me on a journey, both literally and figuratively, to find a way to reframe my view of how social factors produce disability. I knew I needed to get over my personal fixation on the psychological factors that create psychosomatic disorders and look more closely at the social elephant in the room. Of course, examining a person's community to find the cause of an illness is often seen as trying to attribute blame, and therein lay the challenge. I knew I had to bring society and biology together and show them as the integrally linked codependents that they are. Dr Olssen's response to my sociocultural view of resignation syndrome was clear: if you can't say what is happening in the brain, nobody will care.

A doctor's office is a contained, official place that only allows in whispers of the patient's social world. To achieve my goal, I would have to leave my office more than just this once and put aside my Western medical conventions so that I could meet brains *in the wild*. Doctors examine people in scanners, under microscopes and in sterile clinics, but people live in a complex social world. If politics, the media, folklore, social circumstances, healthcare services and life experience had contributed to resignation syndrome, what were they contributing to my patients' disabilities?

As I write this, it has been over a year since I met Nola and Helan, and I have spent a great deal of time listening to the stories of people with similar medical problems in the hope of gaining a better understanding of the various ways in which a person's environment can contribute, without them even realizing it, to the illnesses they suffer. In the meantime, neither Nola nor Helan has recovered. Their asylum status has not been approved and they both remain bed-bound. I visited them as a neurologist, but the more I think about them, and the more I have learned, the less I see their problem as neurological or even medical. That is almost certainly why I felt so useless, standing there in their bedroom. Resignation syndrome is a language that I haven't yet learned to speak. It exists to allow the girls to tell their story. Without it, they would be voiceless.

2

Crazy

Culture: *The ideas, customs and social behaviour of a particular people or society.*

It was sunny again, a warm January day in Port Arthur, Texas. The staff in the motel remarked on the cold, but by my Irish standards it felt more like spring, breaking into summer. I spent the morning wandering through quiet streets, the only pedestrian in a town with few sidewalks. I was looking for somewhere to eat, but could only find squat, residential houses. Most stood detached, each with its own garden, but without dividing fences. The residential area came to an end at a wide highway. I could see drive-in fast-food places on the other side. The reckless jaywalker in me wanted to make a run for it, but there were just enough pickup trucks zipping by to stop me. I couldn't find a crossing, so in the end I bought snacks and coffee in a service station instead. The man behind the counter told me he liked my shoes – London brogues – and then asked where he could get a pair, since he had just got married and he would like to get his new wife some shoes just like mine. In the few minutes it took for me to pay, it felt like he had told me his life story. When he mentioned my shoes again, I wondered for a moment if he wanted me to take them off and give them to him. Then I remembered that this was Texas, where everybody was friendlier than I was used to, and he was only making small talk. Since I arrived, I'd been engaged in conversation by everybody I'd met, each of whom, whatever their age, had called me 'ma'am'.

Even at the airport, the immigration officer had kept me talking. I haven't always enjoyed going through immigration in US airports, but Texas seemed to want me to know I was welcome. When the immigration officer asked me the purpose of my visit, I told him I had come to interview the Miskito of Nicaragua who lived in the region. He looked confused, so I explained that the Miskitos were indigenous to Nicaragua, but that I couldn't visit them there as the country had experienced an upsurge of political turmoil that made it unsafe. So, here I was, in safe and hospitable Texas instead.

'The largest number of Miskitos outside Nicaragua live in Port Arthur,' I told him.

'I knew we had a lot, but I didn't think it was that bad,' he said, smiling and handing me my passport.

As I walked away, I realized he thought I was talking about mosquitos.

The Miskito are the indigenous tribes from the Mosquito Coast, which stretches the full length of Nicaragua and into Honduras. I had learned about them in a news article headlined, *Mass Hysteria breaks out in Central America*. Forty-three people in three communities had been struck down with tremors, difficulty breathing, trance-like states and convulsions. The outbreak was attributed to a condition known as grisi siknis. The *Diagnostic and Statistical Manual* (*DSM*), used by the American Psychiatric Association to classify mental disorders, refers to grisi siknis as a 'cultural concept of distress', meaning it has psychiatric and somatic (physical) symptoms that are uniquely seen within a specific culture or society. It is also sometimes called a folk illness. The communities in which folk illnesses occur usually have their own explanations for them and employ specific traditional remedies to treat them. The story of grisi siknis belongs to the Miskito.

The anthropologist Philip Dennis first described grisi siknis in detail after he came across it in the 1970s. In Dennis' words, those affected 'lost their senses'. It often began innocuously enough, with common symptoms like a headache, tiredness and dizziness, but, as it took hold, it led to the apparently irrational behaviour, convulsions and hallucinations that defined the condition. The sufferers became agitated and aggressive. They swung machetes and threatened to harm others, but were more likely to harm themselves. Some who had been affected were said to carry the scars of grisi siknis for the rest of their lives.

Central to the disorder was a visual hallucination: the sufferers were visited by a frightening stranger – a dark figure, usually wearing a hat – who came to carry the victim away. In the accounts given to Dennis, this figure was referred to as the devil. The form the devil took was different for everybody; it was a personalized hallucination. For women, it was usually a man; for men, a woman. The content of the hallucinations often related to the lives of the people living in the region at the time. For example, in one historical account, a woman reported seeing a black man on a ship, coming for her – this was an era when black traders arriving by boat from the Caribbean were a familiar sight. In another report, a man described seeing a white woman who arrived in a taxi. For another, the devil came on horseback. In a cold reading of Dennis' account of the disorder, these visions do not sound especially frightening, but they were threatening enough that the people who saw them wanted to escape. To do so, they typically ran into the jungle or the mangrove swamp, and had to be restrained for their own safety.

Dennis' description of grisi siknis was filled with eroticism. Often the devil was intent on having sex with the person experiencing the hallucination. Younger women were said to be more susceptible to grisi siknis because the devil preferred them. Dennis reported that some girls became giggly and titillated when recounting what had happened to them. Meanwhile, unaffected male villagers became intimately involved in the girls' drama. They excitedly pursued the victims, fearing that if they reached water they would drown themselves, and believing that it was their responsibility to save them. In a darker version of the legend, it was said that girls caught by villagers risked being subjected to gang rape. Families who restrained their daughters did so in part to prevent that possibility.

Reading about grisi siknis for the first time, it felt like the stuff of ghost stories, full of magical elements, folklore, heroes and villains. Dennis' descriptions were fifty years old and I wondered if the new outbreak I had read about was a singular throwback to the distant past. A quick Internet search said it was not. Grisi siknis was alive and kicking in the twenty-first century.

2003 – *Nicaragua village in the grip of madness.*

2009 – *Nicaragua's supernatural epidemic.*

2019 – *Mystery mental illness affecting girls in Nicaragua.*

These were only the English headlines. There were dozens more in Spanish. With the need to learn more about the role of culture in shaping expressions of distress, I contacted a scientist who had written a great deal about grisi siknis, and she, in turn, introduced me to Thomas, a member of the Miskito community living in the US. Six months later, I was inadvertently misleading Texan immigration officials into thinking that Texas had one of the biggest mosquito populations in the world.

I had met Thomas two days before, in Austin, and we had driven together to Port Arthur. A man in his thirties, he was born in Nicaragua, but had lived in Texas for a long time. He was embedded in the close-knit Miskito community of Port Arthur and he had agreed to introduce me to them. He was trilingual, speaking fluent Miskito, Spanish and English, so he would be both guide and interpreter. Our long car journey between cities had given me ample time to learn how proud he was of the place he came from. He enjoyed his life in the USA, but dreamed of returning to a house among the trees in rural Nicaragua. As an Irish person settled in England, I understood his mindset; I am Irish to my core, but I identify as a devoted Londoner too. On different days, I could put either one of those identities first. I think that those who have never felt obliged to emigrate find it hard to understand how a place in which you no longer live can still feel like home.

Thomas picked me up from my motel after I'd finished my service-station breakfast. He was neatly but casually dressed and wearing a baseball cap. As we drove across town together in his truck, he set the scene of our upcoming meetings. Port Arthur is an oil town, refineries dot the seascape like imposing metal triffids. Even between the pleasant timber houses of the residential areas, there are small industrial plants. Thomas told me that most of the people he would introduce me to had come to Port Arthur as specialists, to work in the refinery. I imagined it helped their decision to emigrate to know that there was a large Miskito community in town.

'Have you ever seen grisi siknis?' I asked him.

'Yes.' He nodded. 'When I was still in school, it happened to some people I knew. The school was closed and we were sent home. They locked the ones who had it in a classroom.'

'They locked them in?'

'Yes, they have to do that, or they run into the jungle. These people will tell you,' he said, pulling the truck up in front of a house that had half a dozen cars parked out front. It was a typical single-storey home, painted in muted colours, but the cars made it look as if it was about to host a party. Music escaped into the street through an open window.

Ignacio, a stocky man in his forties, appeared at the front door, greeted us warmly and ushered us inside to sit at a big rectangular table. Instinctively, I took a corner seat with my back to the wall. I wanted to be able to see everything and absorb the atmosphere, but I also felt somewhat compelled to hide. As I sat down, I knew I was being watched. A few people were at the table already and others milled around the kitchen and living room. I was given quick and confusing introductions. A Christmas tree in the corner was the reminder I needed that the new year had barely started. Cooking smells filled the house and a huge sound system throbbed with drum-heavy music.

'Nicaragua music. You like it?' Ignacio asked.

This was the home of Ignacio and Maria. They had lived in the US for more than thirty years. At Thomas' request, they had gathered a group to meet me. Looking at the men and women around the table, what was immediately obvious was that being Miskito was not a question of race. A Miskito person couldn't be recognized by skin colour or style of dress or general appearance. Everyone in the room looked different.

Most were in their forties or older. Over the course of the day, I would learn that they had all been born in and spent their formative years on the Mosquito Coast, but their place of birth was only part of what gave them their cultural identity. Some of them, like Thomas, identified strongly as Miskito, while others considered themselves North American.

Like the majority of the population of Nicaragua, Miskito are mostly mixed race, largely made up of white European and Native American heritage, but with many other variations. The largest European contribution is Spanish. There is also a heavy influence from the North and West African people who were brought to the Caribbean as slaves. Britain, Germany, Italy, China and many more nations contribute to the ethnic diversity of Nicaraguan people. Native indigenous tribes make up only 5 per cent of the population. Although the Miskito are descended from indigenous South American tribes, the present-day Miskito person has multiple other cultural influences. I was neatly reminded of that as I sat in Maria's kitchen, watching her cook rice and peas, listening to Nicaraguan drum music, surrounded by Christmas paraphernalia.

'Some people say there are no pure Miskito left,' Thomas had told me, 'but I think there are a few in the remoter villages.'

Thomas came from German, English, black Caribbean and indigenous heritage. Everybody around that table was a different mix of white European, black African and Native American ancestry. My visit took place in January 2019, just as thousands of migrants were travelling northwards through Central America, hoping to seek asylum in the US. Every news broadcast featured Donald Trump's plan to build a wall at the Mexican border. It was an interesting time to be with these people. Their native culture had been subsumed centuries before, and their religious beliefs and national language had changed irrevocably. The Miskito identity was very important to them, but they had incorporated new influences into it and I could detect no bitterness on that account. Places change over time and people adapt. Later, I would meet their children, those born in the USA, and see how these members of the next generation of Miskito were shaping their beliefs to fit their lives.

Maria was in her forties, like her husband, and was cooking furiously, alternately stirring a big pot of spiced chicken and a pan of yellow rice. Her mother worked beside her, chopping vegetables and using a strange contraption to slice plantain into long thin strips for deep-frying. Every now and then, somebody knocked on the front door, but no one waited for an answer. They just allowed a second or two to pass and then let themselves in. Some came directly to sit at the table, while others, whose priority was Maria's cooking, went straight to her with Tupperware to be filled. It seemed Maria had a reputation as a cook and was preparing Saturday lunch for the whole neighbourhood.

The kitchen and sitting room were two distinct areas, but without a dividing wall. The floor was wooden. It looked brand new, except it stopped with an unfinished appearance at the entrance to a utility room, where it gave way to buckled, stained linoleum.

'We're still repairing after the storm,' Ignacio told me when he saw me looking. 'The water was up here.' He indicated a level just above his knee.

Two years before, Hurricane Harvey had flooded every house in the neighbourhood. It was the first flood that Ignacio and Maria had seen in thirty years of living in Port Arthur. As I'd walked around the neighbourhood earlier, I had found it unusually low level, flat and with few buildings higher than one storey.

Driving along the coast to Port Arthur, I'd seen that most of the houses stood on concrete stilts. They were so tall, I couldn't quite believe it; I'd asked to stop the car, just so I could stare up at them. Most had three storeys of steep concrete stairs before the house even began. But it seemed that protection against global warming was yet to arrive in Port Arthur.

In the Mosquito Coast, where this community of people were born, houses have long been built on stilts to protect them from the heavy tropical rains, which existed long before climate change came to threaten Texas. Many years ago, and before I knew that grisi siknis even existed, I'd travelled through the Mosquito Coast as a backpacker. In my memory, Miskito houses were ramshackle, wildly different to these Texan homes. They were embedded in the trees and red soil, constructed from local timber, but with gaps in the walls and patched corrugated roofs which let in the constant roaring noise of insects.

Maria and Ignacio's home was all-American, modern and decorated for the season. I waited until Maria had served lunch and we had all finished eating before I tentatively asked the people around the table whether they had any personal experience of grisi siknis. I knew something of the folklore of the disorder, but still held on to a suspicion that it was a fairy tale, overblown by media reports.

'Yes, I've seen it,' a man opposite me answered. 'Everybody's seen it.'

Around the table, heads nodded. They had all seen somebody with it. Some of them had family members who'd been affected. It was not just a myth sensationalized by news reports. It was something real in their ordinary lives.

Chewing on a toothpick, Anthony, a man in his forties, took early control of the conversation.

'I saw it plenty of times. Plenty,' he said, nodding sagely. 'I was born in the seventies, in Puerto Cabezas, but aged nine I moved to a small place called Awastara. People I knew when I was there had it. They said they saw a leprechaun who gave it to them.'

'A leprechaun!' I laughed, and they all laughed with me. I thought I had left the land of leprechauns behind me nearly twenty years before. 'But leprechauns are Irish, aren't they?' I said.

'A leprechaun is a small man,' Anthony said, correcting me. Several other people nodded in agreement. 'They are small men that live among the cows. If you see a large group of cows together, then there could be a leprechaun there. A leprechaun can make a person crazy – the kind of crazy that makes people climb coconut trees.'

'Really, it's true,' Ignacio agreed. 'Even if a person can't climb a tree when they're well, they climb like a monkey when they have the sickness, and the small man gives them the sickness.'

'And he makes them do other things,' Anthony added. 'A lot of times, they break glass and then they eat the pieces.'

'Yes,' Maria agreed. 'I saw a girl eat glass like it was nothing.'

Once more, the whole table was in agreement; they had all heard or seen strange things. But the pieces of the story were disjointed. I found myself musing on the leprechauns. Already the story had departed from what I had expected and made me wonder just how many cultural influences had blended to create grisi siknis. I asked if Awastara had ever had Irish missionaries.

'We have a Moravian church,' Anthony said.

The Moravian Church originated in Europe and is one of the oldest Protestant denominations. If there was a Moravian church in Ireland, I had not encountered it. All the people at Ignacio and Maria's were devout Christians and they had set up their own Moravian church in Port Arthur.

'It's not always a small man that visits,' said Thomas, who could see that this had become a stumbling block for me.

'No. It could be a dead person,' Anthony confirmed. 'Maybe it's somebody they know, who comes back. It's different types of spirits. It makes the sick person run around. The spirit, whoever it is, enters the body and makes the person act crazy. Then, the spirit, it passes from young lady to young lady. You get too close, you get it too.'

'But it's not just girls?'

'No. It's boys too. It's old people.'

'No, old people don't get it,' Ignacio countered. 'Just young people, especially girls. You're too old to get it,' he said, indicating me.

I started laughing again and a small argument ensued in Miskito on the subject of who could be affected. I couldn't decide if it was better to be so old as to be invulnerable, or to be considered young enough to fall victim. In the end, they agreed that it was mostly young people, but some of them had seen it in the elderly too. Accounts in the literature and media indicate that it is predominantly a disorder of teenage girls, but certainly there have also been cases in teenage boys and some older people. Once a woman got married and started to have children, she was less likely to get the sickness.

'You know grisi siknis, it means "crazy sickness",' Ignacio said.

When I had learned this first in my reading about the disorder, I was surprised that I hadn't realized it immediately. Miskito might be an indigenous native language, but it has been corrupted over the years by English and has many English loan words. It is written phonetically, making it easy to read and easy for a non-speaker to guess at the spelling of words. The bluntness of the name surprised me. There are not many good interpretations of the word 'crazy'. Yet, when people around the table used that very word repeatedly in their description of the disorder, it did not feel judgemental. When the person with grisi siknis began to behave strangely, it was because a spirit had entered them or corrupted them or frightened them. Their behaviour was not a reflection on them; although they were behaving as if they were crazy, it was not their fault.

'Many times it comes because somebody has read the black book.'

'The black book?' I asked.

'The black book is a devil-worship book,' one person told me. It was also used in voodoo, they said. Once again, there was much agreement around the table on the part played by the black book in a person falling ill.

'If you gave me the book, I wouldn't read from it. I wouldn't even touch it,' Thomas told me. 'Sometimes the people read from it because they're just curious, and then people start to get sick.'

'Men use the black book to attract women,' Maria said.

Here was something familiar. The descriptions they had given me so far were different from those that Philip Dennis had given in his study of grisi siknis. But, with Maria's comment, the idea of it being a disorder related to sexuality and eroticism had entered the conversation.

'A taxi driver in Puerto Cabezas told me that he could get any young girl he wanted with the black book,' Anthony said.

'A man did it to my sister,' said Lucia, a glamorous woman in her fifties, who had been sitting quietly listening until then. 'It's why I left. I was afraid he would do it to me.'

Lucia's sister was only sixteen when she developed grisi siknis. At first, she had complained only of dizziness, but then her behaviour grew erratic and she became frenzied.

'She was a small girl, but she got so strong it took seven men to hold her down. I saw her pry up the wooden floorboards with just her fingers. Just her fingers!'

That the sufferers became super-strong was a recurrent theme of the stories I was told, and common threads ran through all the accounts. A person, usually a small man, visited and brought the sickness. The black book was mentioned often. I asked where I could see the book, but nobody at the table wanted to admit to knowing that. They repeatedly emphasized that grisi siknis was done to a person by somebody. It often seemed to be the case that older men did something to the younger women to make them sick. Convulsions, foaming at the mouth, *crazy* behaviour, ripping off their clothes, running manically, hyperventilating, and breaking and eating glass were the symptoms and signs most consistently mentioned.

I wondered about the idea that older men perpetrated grisi siknis on younger women.

'Did the man do something physical to hurt your sister?' I asked Lucia.

I was having difficulty suppressing my Westernized medical interpretation of this worrying part of the story. How real or imagined was the threat the girl faced? Lucia assured me that, whatever the man had done, it was from a distance. He had physically harmed her sister, but only through invoking magic. The girl developed convulsions and confusion. The family kept her locked in and restrained her by tying her with rope. If they hadn't, they were sure she would have hurt herself. Lucia's sister had insisted that something was inside her stomach and was trying to get out.

'She got sick and a ball of hair came from her stomach,' Lucia said, clearly still upset at the memory.

The family was terrified and took the girl to both a conventional doctor and a traditional healer. The first did nothing, and the second gave the family herbs and advised them to bathe her with them. The villagers rallied around the family and confronted the man.

'The man said he had glass hands, when they accused him,' Lucia told me.

'Glass hands?'

'Clean, so you can see through them.'

In other words, innocent. The villagers didn't believe him, so they drove him out of the village and told him he should never come back. But, even at a distance, he could still harm anyone he wanted, so Lucia had moved to the USA in fear that she would be next. Grisi siknis only happens in the Mosquito Coast area. It doesn't affect Miskito in bigger Nicaraguan cities or those outside the country. It occurs mostly in small villages. Lucia escaped and her sister eventually recovered, but much later they heard that the man, who had fled to Jamaica after his banishing, had died. Nobody knew of what.

I asked the thing I had been longing to ask: 'Do you think it's possible that this illness is psychological in some way? If I'm honest, if somebody had the behaviour you describe in England, they would probably be referred to a psychiatrist. Does that make any sense to you as an explanation?'

Everybody shook their heads in a definite *no*. A few laughed.

Ignacio spoke for them: 'The crazy sickness is not like a normal thing. They have seizures. They are so strong – you can't even imagine it. Five men can't handle a small girl. They bite. They pinch. They grab you and you can't get away.'

'I saw it in a school,' Jaylene said. A woman in her thirties, she had been sitting quietly listening, smiling widely the whole time. 'It looked like epilepsy. It's as real as that.'

Jaylene had worked for the police in Nicaragua. She had been called to a school to help contain an outbreak of grisi siknis in which almost an entire class had been affected in a single day. The police were asked to stop the children running away, while the teachers waited for parents and healers to arrive. The epidemic had spread quickly. It spread by touch, but also, if one child said the name of another, they would be the next to be affected. Outbreaks of grisi siknis in schools are common, I learned.

'So how do they stop it?'

'They tie them up and lock them in. Then they wash them in herbs that the healer gives.'

'Is grisi siknis the same sort of illness as epilepsy? Is it considered a brain disease?' I asked. I was wondering where it fell in categories of illness, from their perspective.

'It's not epilepsy. It looks like it, but it's not that. The healers can cure it. It's medical, but it's due to black magic, so it's different too.'

Western medicine does not leave much leeway when it comes to the classification of illness, and, as a result, I was trying to crowbar grisi siknis into a category that I could understand. In my clinic, I see many people with seizures just like those described. It is very common for relatives to say that a person with dissociative (psychosomatic) seizures becomes very strong and has to be pinned down. A slave to my training, I was trying to find a word for what they were describing that would fit my medical lexicon. But no such word existed and, when I asked them to classify the illness, they struggled to answer my question. Eventually, Jaylene explained to me that it was not as simple as saying it was a disease like other diseases. It was a medical problem, but not the same as asthma or cancer or diabetes. What was absolutely clear was their expectation that, if you took a person with grisi siknis to a hospital doctor, even in Nicaragua, they would say that it was a psychological problem, but wouldn't offer any treatment. In the view of the people around the table, that made the doctor's opinion useless. But the local healer's treatment was curative, so, when someone succumbed to the condition, the family typically called the healer and the pastor.

Maria went to a cupboard and got a small plastic bottle filled with clear liquid. She opened it; it smelled of overripe fruit.

'Florida water – this is part of the treatment.' Florida water is a perfume that some say purifies the spirit.

She then went to the utility room and reappeared with a small plastic bag filled with a blue powder that she said could cleanse a person of the sickness. She also brought a container shaped like the Virgin Mary, full of holy water. When I was growing up, these featured in every Irish home I knew. My family had been given one by somebody who had been on pilgrimage to Lourdes, where the Virgin Mary was said to have appeared. Maria told me that holy water was mixed with Florida water, along with the blue powder, lemons, garlic and a variety of herbs. The grisi siknis sufferer was then doused in the concoction. Many get better very quickly with this treatment. Sometimes the illness lasts months, but the vast majority recover completely.

'The devil is looking to be warm, so he enters the girls,' Anthony explained.

Jaylene's husband, Mario, turned to me. 'Do you believe in the devil?' he asked.

I thought for a moment. 'I don't believe there is a single person or entity that is the devil,' I said eventually, then asked, 'Do you believe in him?'

He shook his head. 'I have never seen him, so I have no proof of him.'

'But you haven't seen God either, and you believe in Him.'

I knew that Mario was a devout member of the Moravian Church.

'No, I have never seen Him. But I feel Him.' He smiled at how obvious this was, and perhaps with a degree of sympathy too. He knew that I couldn't feel God.

'Do you have any thoughts about why young girls are more likely to be affected by grisi siknis?' I asked him.

'I don't know . . . but I think maybe the girls are weak and grisi siknis makes them strong.'

I thought so too.

'What is life like for young people in the Mosquito Coast?' I asked.

'It's not like this,' he said, indicating everybody and everything in the room. 'There, the girls go to school and then they come home. They don't go out. They don't have freedom. Here, they go where they please.'

'Do you think it brings them attention that they can't get in other ways?' I asked.

'Maybe it does, maybe it don't.'

What was clear, as the conversation progressed, was that the spiritual beliefs of the Miskito were central to grisi siknis. Most of the group had lived in the US longer than they had lived in Nicaragua, but the language, music and religious influences of their childhood held firm. What was also inescapable was that grisi siknis brought a community response into force. Treatments that might seem cruel or strange in European or North American society were actually very effective. They tied up the teenagers and pinned them down. They threw buckets of the healing potion over them and prayed for them.

Listening to each person tell of the sister or friend who had been affected, I saw that there was very little shame attached to the disorder – among this group, at least. They had talked so freely about it, knowing that I would likely think their stories odd, but still trying to make me see they weren't mad or illogical or unusually superstitious. These were all educated people and skilled workers. They depended on conventional medicine for their healthcare as much as I did. Grisi siknis was ordinary to them and coexisted easily with their more Westernized selves.

'You want this?' Maria offered me the bag of blue powder.

'I'm not sure.' I hesitated; I was thinking of the implications of trying to board a plane with a bag of powder I couldn't identify.

Maria read my expression and started to laugh. 'It's okay – if you don't need it for grisi siknis, you can wash your clothes with it too!' she said.

That evening, they took me to church and invited me to stand before the pulpit to talk about my work, which was both an honour and a shock. I had not expected to have to speak. They were a kind, receptive audience, and I tried not to let my atheism show.

After we had said goodbye, I found my way to a roadside fast-food restaurant where I could sit and absorb what they had told me as I watched the Texan world go by. Sipping my coffee and making notes, I thought of the blue powder, regretting that I hadn't taken it. That opportunity lost, I logged onto the Internet and bought some Florida water instead, which was waiting for me when I got back to England. I haven't used it yet. I am keeping it in reserve until I feel in need of the luck and prosperity promised by the website that sold it to me.

During my trip to Port Arthur, Mario said to me, 'Miskito – if they have four stories, they will only tell you three.' Somatic (bodily) symptoms that have a psychosocial cause have something in common with that description. There is a story being told by the symptoms, but, while the physical manifestations may be explicit, the complex factors that led to them are not so easy to see. The Miskito were warm and welcoming, and willing to be asked and to answer anything, but, to understand the disorder in a more academic way, I met a doctor of anthropology who had studied grisi siknis and had written a thesis about it. She had lived in a village in Nicaragua, but she was Italian by birth and now worked in Paris, which is where I met her.

Paris, of course, is deeply connected with the story of hysteria. It was the home of the nineteenth-century neurologist Jean-Martin Charcot and his 'hysterical circus'. Charcot had spent decades trying to understand psychosomatic disorders (then called hysteria) and held weekly lectures at the Salpêtrière Hospital, at which he used his patients to demonstrate the seizures and bizarre behaviours that were characteristic of the condition. His audience comprised doctors, artists and members of the public – Freud among them.

I was very happy to have an excuse to go back to Paris, where I once lived for a wonderful, but all too brief period of time. I arranged to meet Dr Maddalena Canna in a coffee shop not very far from the Salpêtrière. Over the din of clinking cups and Sunday traffic, she told me about the time she had spent in the Mosquito Coast studying grisi siknis. While there, she had been hosted by a family and had learned to speak and write Miskito. She was fully immersed in the community and retained close ties with the people there. For a year, the family she lived with listened for reports of grisi siknis and took her to the sites of fresh outbreaks. She had agreed to describe her experience to me, to give a more objective view of the condition. But it was a heartfelt account, too, since those she had lived with in Nicaragua had become like a second family to her. She had even adopted a child from the community.

'The Miskito's classification of illness is very complex,' she told me from the outset. 'Europeans who hear about demons and spirits sometimes assume that means that they are a naive people. That is not how it is at all. Often, they have insight. They are in control of it, to a degree. Grisi siknis is a way to exteriorize conflict.'

Madda's comments about outsiders' reactions to stories about demons certainly applied to me. The first time I heard of the grisi siknis, I referred to the beliefs surrounding it as superstitions. I grew up in Catholic Ireland. Throughout my childhood and into my early twenties, I went to Mass every Sunday, and every day during Lent. My mother made us kneel in the evening for family prayers, albeit in fits and starts of devotion. Why was it that we always ate fish on Fridays? For Catholic reasons that I forget. When I became more independent in the world, I lost my faith, but I think it is worth noting that I have never referred to the spiritual beliefs I was brought up with as superstition.

I started by asking Madda how the community classified the disorder. Was it a disease, a psychological disorder or something entirely distinct? In conventional Western medicine, categories matter a great deal. They decide what sort of doctor you will see. They are needed for treatment protocols and research and for charging patients. I was still struggling with the need to fit grisi siknis into a schema I recognized.

'I would say there is no consensus on that,' Madda said. 'Most people would say that it is a disease – but a disease that is caused by a spirit. Many refer to the spirit as the "*duende*". The explanations, like the condition, are very changeable, but in general it is seen as a biological disorder induced by something spiritual. That is not to say that there are not those who regard it as a psychological condition – of course there are. It is a fluid condition, with a great deal of paradox. I have even met one person who said that the evil spirit who caused grisi siknis was itself a psychologist!'

The average age of onset for the condition is sixteen, she told me. It typically occurs during the transition from the teen years to adulthood. Philip Dennis was the first person to write about it, in 1981, but texts dating back to the seventeenth century describe a similar condition, albeit with different names and attributions. I asked Madda if she agreed with Dennis' description of a sexually charged and exciting phenomenon.

'It is an eroticized condition, there is no doubt,' she told me. That said, though, she believed reports of those affected being raped by those chasing them to be the stuff of folklore. 'The narrative is sexualized and the evil spirit that attacks the people is often a seducer. But that is part of the imagery of the disorder, rather than representing a real occurrence.'

Personalized visual hallucinations are fundamental to grisi siknis. In keeping with the theme of seduction, women often see an attractive man, while men are more likely to see a desirable female demon.

'For example, one woman reported seeing a handsome man, carrying a sword, coming towards her,' Madda explained. 'The disorder is linked to a conflict between the moral standards of the community and awakening sexuality. Desire is seen as a satanic attack.'

In the Miskito community, young girls are subject to a lot of attention from men. Often, those men are older. One teenage girl may have several men trying to pressure her into a relationship. Sometimes, the interest paid by men is welcome and exciting for the girl. However, it starkly clashes with the moral standards of a community heavily influenced by conservative branches of the Christian Church. The girls might feel threatened or titillated by the attention, but, either way, they couldn't act on it.

Grisi siknis was also used by some men, who told girls that they would curse them with it unless they agreed to sex. In those cases, the hallucinated figure that came for the girl usually took the form of the man who had carried out the sorcery. And, of course, it is not only a female condition. Boys were occasionally also prey to seducers – usually, in their case, women.

'Being affected by grisi siknis means that you are attractive,' Madda told me, 'so it is exciting, while also being embarrassing. But it is not just about sexuality; it is related to conflict in general. One girl I met told me that the spirit who came to her was in the form of a baby. That girl had assisted in an illegal abortion. Having grisi siknis gave her the opportunity to act out how she felt about that. It is a very sophisticated process.'

'Is it stigmatized?' I asked. 'Is it shameful to be affected?'

'Shameful?' She paused. 'A little. It can mean you are weak, and that you have succumbed to a demon. Maybe it means you have had intercourse with a demon. But is it stigmatized? No. The presence of the demon allows the people to escape stigma. A spirit affecting a person means they are not mad.'

The opposite, then, of how it would be in Europe or North America, where reporting seeing a demon would be thought to indicate a mental-health problem. The Miskito people I met in Texas certainly didn't associate the condition with a psychological disorder. They also gave a somewhat different account to the one I had read about in textbooks.

'The symptoms are very plastic,' Madda told me. 'There is a lot of scope for projection from one person to the next.'

Grisi siknis is highly contagious and is shaped by suggestion and expectation. It can also spread like wildfire. Schoolchildren are especially vulnerable. Typically, it starts with a single child, and then, one by one, almost a whole school can be affected. Madda had witnessed one such outbreak.

'It looked like a panic attack,' she said. 'First, their heart starts beating very fast, then they breathe very heavily, then they collapse. They say it is like falling into a nightmare.'

'What do the schools do when this happens?'

'Usually the school director closes the school and the children are restrained.'

Again, she described the excitement of the event. The convulsing movements looked sexual, and bystanders from the village often watched the girls, making grisi siknis a social spectacle. The children were protected by their relatives. Usually a shaman was called, although some schools also called a medical doctor and a pastor.

The type of help requested largely depended on what was available. Many Miskito villages are very isolated, making the local shaman the only viable option. When I asked Mario about the village in which he grew up, he told me that, when he was a child, his village had no electricity. It had only been installed in recent years, which had turned out to be both good and bad. In Mario's own pithy words: 'When people told us about the miracle of electricity, they forgot to tell us about electricity bills.' In a similar way, the arrival of modern medicine has come with benefits and problems, resulting in a tug of war between new and traditional methods. Because Western medicine is generally portrayed as being superior, both worldwide and within the Miskito community, some grisi siknis sufferers prefer seeing a doctor to a shaman, if that option is available. But, in hospital, they find themselves treated with benzodiazepines and sometimes even epilepsy drugs. These treatments don't work, while shamanism is largely successful.

'Also, if they are given injections in hospital, the seizures can get worse,' Madda told me. 'Injection is seen as penetration and is associated with a sexual act. People wear amulets to protect themselves against demons and they believe that injection with a needle will destroy the power of the amulet and could even kill the person.'

Traditional healing methods, on the other hand, have been very successful in treating grisi siknis. The vast majority of sufferers get better. A few develop chronic or recurrent symptoms, but that is not usual. The ongoing reliance on traditional healing methods has meant that local hospitals have had to find ways for old and new treatments to work together. In fact, Nicaraguan law requires the incorporation of cultural concepts of illness into healthcare models and encourages cooperation between healers and biomedical practitioners. People have trust in shamans and their intervention works. As Madda put it, 'The shaman's treatment is symbolic. It is really a psychological intervention and it works better than benzodiazepine.'

Grisi siknis is influenced by the stories told about it. During Madda's stay with the Miskito community, she became part of the narrative. In the early days of interviewing people about the condition, she found some were reluctant to describe the experience in detail. She hit upon the idea of asking them to draw a picture representing their experience instead. They resisted at first; pictures of the *duende* are said to have magical properties and could make other people sick. However, some agreed, and one day a mother remarked that the picture had the effect of drawing the demon out of her child. This discovery moved Madda from the role of ethnographer to shaman – a role she was at first reluctant to accept.

'I told them I had no medical training. I couldn't be responsible for that.'

But when some people were cured by the act of drawing, she continued with the practice. She showed me some of the pictures. Most were of strange, stick-thin, faceless figures, all wearing hats. Some had oddly shaped or absent thumbs. The *duende* wears a hat, has only four fingers, and his or her face cannot be shown because it is believed that puts those who see it at risk of catching grisi siknis. One picture showed an armadillo, an animal associated with the demon. Another displayed the spirit with a stomach swollen with blood. Stomach parasites are common in Nicaragua, and stomach complaints can be a trigger for grisi siknis: the sufferer catches the infection and mistakes it for the onset of grisi siknis, which can also begin with abdominal pains. The assumption that they had contracted grisi siknis could cause a person to anticipate seizures, and that could trigger dissociation, causing dizziness and ultimately fulfilling the person's expectation. So a medical illness could also be the inciting incident for the syndrome; it does not require psychological distress or conflict to be present.

'Would you like to see a video?' Madda asked, pulling her laptop from her bag and turning the screen towards me, away from the rest of the coffee drinkers.

The details of the video were hard to make out, at first. The picture was dark and there were a lot of people moving around in shot. The rumble of voices in the Parisian cafe was just about loud enough to cover up the grunting and crying sounds coming from the computer. As my eyes adjusted to the scene, I could see that a young woman was being carried by a crowd of people. The camera only caught glimpses of her through all the people trying to keep her still. Her head was thrown back and she was struggling. They supported her in mid-air, as if trying to transport her somewhere. A woman pulled her shirt down, tying the tails in a knot, in what seemed to be an attempt to keep the girl covered. The girl's head thrashed from side to side. The people who held her seemed to be fighting to control her limbs.

'Have you seen anything like this before?' Madda asked.

I had seen something like what was in the video a thousand times before, probably more. It looked very much like the dissociative (psychosomatic) seizures I see every day of my working life. The families of my patients often bring me videos of the attacks. In them, I see people pinned down by their loved ones, just like this. I hear frightened screaming and crying, and sometimes praying. The videos I am given have city backdrops, they take place in cars and in modern living rooms; but for these superficial differences, the scenes they show are very similar to the one Madda was showing me. My patients' seizures last for several minutes, during which time they stop and then restart repeatedly. The girl in the video did the same. In between convulsions, she froze in a catatonic pose.

'When I took this video, the people told me I would catch grisi siknis very soon because of it,' Madda said. 'An artist once drew a painting of the *duende* and the community objected until he was forced to change the painting to make it more subtle. Imagery and storytelling are very important to how the disorder takes hold. They said that the drawings would make me sick. Now there are lots of newspaper articles and images on the Internet about the disorder, and this is yet another source of paradox and conflict that the community has to tolerate.'

'Did you think you would catch it?' I asked.

'I wanted to so badly.' She laughed. 'I wanted to know what it felt like. Once, while I lived there, I picked up a stomach parasite and they said that was the start of it!'

'But it didn't happen?'

'I was always looking out for it. But, no, it didn't happen. Although, people still say that I will get it when I stop looking for it.'

'There's still time, then.'

But, to get it, you have to be in the Mosquito Coast. It is a medical disorder that is strongly connected to an ancient tradition, and the spirits are tied to the place, even if the Miskito are not.

Resignation syndrome drew attention to Nola and Helan in a way that a purely psychological expression of distress could not. It is often more efficient and productive to experience and exteriorize psychological distress and conflict as bodily symptoms. But why grisi siknis in Nicaragua and resignation syndrome in Sweden and something else in the UK? Illness is a socially patterned behaviour, far more than people realize. How a person interprets and reacts to bodily changes depends on trends within society, their knowledge, their education, their access to information and their past experience of disease. Personal and societal role models create expectations of health and ill health that are coded in neural substrates. Our brains are wired through experience to respond in a certain way to certain provocation. It is an unconscious process.

At birth, the brain is a blank canvas that is full of possibility. A newborn has more brain cells than an adult. However, the cortex, the grey matter, does very little in a newborn. It has to be brought online, and that happens through experience. After birth, myelin, the insulating layer that forms around nerves, grows along axons, and in doing so promotes the spread of messages between different areas of the brain. Learning is mediated by the development of connections between networks of neurons, so clusters of cells in different parts of the cortex start speaking to each other and storing new information. This is hugely dependent on environmental stimuli. The somatosensory cortex comes online first in response to touch.

The colourful mobile hanging over the bed helps stimulate the visual cortex into action. Thus, a child's immediate social environment is shaping the brain from the day of birth, and even before.

Language is a perfect example of how brain development is socially influenced. We are all born with the capacity to understand and speak any language. As soon as we are exposed to words, our brains strengthen the connections that are required to understand the units of sound that make up the language of our caregivers. As one language is learned, another is lost. To make our brains more efficient, connections are also pruned away. We strengthen the connections we use regularly and lose those we don't. Thus, our capacity to hear and pronounce the components of languages that we are not exposed to diminishes with time. That is why the perfect pronunciation of French semivowels escapes me, as does the rolled 'r' of Spanish; I wasn't exposed to either until I was a teenager.

The brain is a cultured organ. It depends on exposure to learn. Only a small part of learning is conscious. Every day, with every person we meet and every conversation we have and every piece of music we hear, as our brains process that environmental stimulation, they are also changing to accommodate it. Things that only happen once, or experiences that are not emotionally significant, won't leave much of an impression, but repeated or emotionally charged experiences leave indelible marks.

In just the same way, our ideas about illness and health are woven into our brains. Our attitudes to bodily changes, how we interpret and react to symptoms, who we turn to for help, how we explain disease, the treatment models on which we rely are all learned. It is a fluid system, of course. As new influences move in, our brains adapt to accommodate them. Below the level of our conscious awareness, an internal physical feeling is translated into a response by cultural norms. If it's winter in the UK and you feel unusually tired, you might think you're developing the flu – you reach for the vitamin C and paracetamol and take to bed. Somebody else, living somewhere entirely different, will follow that identical feeling of tiredness to a completely different cause and solution. In Madda's words, 'we embody cultural models of illness.'

Embodiment is central to the development of psychosomatic disorders. There is sometimes a misperception that thoughts exist mostly within our heads, with our bodies playing little or no part. Actually, the body is intimately involved in cognition. If I were to ask myself to draw on any emotion, I would immediately feel the physical experience of it in my chest, in my limbs, in my skin. Bringing to mind an evocative memory creates a change in my heart rate, muscle tone, hair follicles. In the theory of embodied cognition, that mind–body interaction works the other way around too. The physical symptom could come first, alerting me to a change in how I am feeling. Butterflies in my stomach, created by autonomic arousal, appear as a reminder of some anxiety in my life. Palpitations tell us we're scared, and blushing alerts us to embarrassment. Thoughts and emotions are felt in and enacted by the body. That is not to say that we are always accurate when we interpret these physical changes. Do those butterflies mean excitement or dread?

We physicalize mood, emotional well-being and even personality. Confident people stand with confidence. A shy person carries themselves differently from an outgoing one. The same person might have a completely different way of sitting when they are feeling down, versus when they're happy. We use the body language of others to predict their opinions, attitudes and moods. We also use physical gestures to communicate, both consciously and unconsciously. We smile to indicate happiness and nod to imply agreement. But, here, we need to be careful, because culture is very influential over body language.

The Indian headshake or head wobble is an example of a culturally determined physical gesture. It has nothing to do with race; it exists only in people who have spent their formative years in a particular cultural environment in the Indian subcontinent. It is a gesture that has meaning to those who speak that cultural language, but which is poorly understood by outsiders. All societies have their own way of expressing themselves physically, either in terms of facial expression, gestures, or just through posture. Junior workers in some countries stride into an office like they own the room, but in other places they shuffle in deferentially. Like verbal language, body language differs from place to place.

We also embody culturally shaped ideas about illness. Like learning a language, we interiorize illness templates; we code them in our brains and then express them physically when triggered to do so. When we experience an unexpected physical sensation, the first thing we do is fit it into a schema that we know. Prototypes that explain somatic symptoms are embedded in our brains. A sore throat means a viral infection. A headache could be a migraine to a person with a family history of migraine, but could mean a brain tumour to somebody whose father had a tumour. And so on. Pathology is a fact independent of the observer, but how one responds to symptoms is drawn from knowledge and experience. Even in the absence of pathology, a single physical symptom could trigger a cascade of further symptoms, drawn from nothing but expectation. As with grisi siknis, the first symptom leads to others according to the familiar narrative of the known condition.

The brain's prototypes for illness are culturally determined.

In the UK, a recurrent stomach upset might make a person think they have irritable bowel syndrome, but that same symptom in the Mosquito Coast could be interpreted as indicating a parasitic infection, or perhaps the onset of grisi siknis. These are each conditioned interpretations of a bodily change that can influence subsequent illness behaviour. The embodiment of these sorts of illness templates can create a cycle of behaviour that leads to functional symptoms.

Culture doesn't only shape our reaction to normal and abnormal bodily changes, it also prescribes the best way to express distress or ask for help. Some problems are hard to articulate, and, for those, somatic symptoms may be a more acceptable way of seeking support or comfort. Physical illness can be a socially choreographed way to let those close to us know we are in need. The symptoms that will work best for this are culturally attuned. In the UK, expressing flu-like symptoms might be an easier way of attracting the care of a loved one or getting permission to take a break from work than having to admit to feeling overwhelmed. Similarly, for an asylum-seeking child in Sweden, resignation syndrome tells people exactly what is needed, with a language that is much more powerful than words. And grisi siknis indicates a particular sort of social conflict, without the need for an explicit declaration. The process of sending out this distress signal is not necessarily a conscious one. The need for help is felt physically through embodiment, and the physical symptoms can be expressed preferentially over the psychological ones because they are a recognized culturally coded message.

The templates in our brains that help us to assess physical changes, and which create models for experiencing distress, also determine who we turn to for help. In the UK, a person with stomach cramps might try managing it alone, or they might turn to a health-food shop, alternative-medicine practitioner or a doctor for help. Western societies' heavy reliance on pharmaceuticals might see a person ask their doctor for medication. They might even garner a referral to a gastroenterologist for invasive investigations. In the remote parts of the Mosquito Coast, where access to modern healthcare is usually more limited, there are fewer treatment routes, making medicalization less likely and spiritual explanations potentially more practical.

Because Western medicine is evidence based and relies on scientific method, it is usually assumed to be superior to other medical traditions. Therefore, it might seem that the person with the greatest choice and easiest access to modern medicine is the luckiest, but we shouldn't assume that is always the case. Western medicine promotes the pathologizing of every bodily change. Once a person has been diagnosed with a bowel condition, it not only enters their medical records, it also enters the individual's unconscious and can therefore affect how that person views their health in the long term, and how they are viewed. Medical conditions in the Western formulation have a tendency to become chronic. They are labels that one can never completely shake. On the other hand, there is something beautiful in grisi siknis. One does not have to subscribe to a belief in demons to admire it. It is an illness with a set course that leads to recovery. It can be a dramatic illness, so it shouldn't be underestimated, but it is noteworthy that traditional treatments are usually curative. Grisi siknis is an illness template with a beginning, middle and end.

The unconscious embodiment of cultural models of distress gives people the means to act out conflict and to ask for help in a way that will attract the right sort of support, the sort that is free of judgement and stigma. Possibly the most fascinating and useful thing to notice about grisi siknis is the way that it attracts a community response. As Madda emphasized to me, this is a highly sophisticated and effective way to deal with conflict. It rallies the community and keeps the affected person in the group. In stark contrast, people with psychosomatic and functional disorders in the UK and US often feel abandoned and isolated, ostracized by the community.

That we in the West are less likely to use spiritual explanations to explain illness should not tempt us to think that the Miskito's response is less logical, more fantastical than ours. We have plenty of our own folk illness beliefs and traditional healing methods, but because they are ours they don't strike us as especially odd. Chicken soup for a viral infection is our blue powder and Florida water. A person won't catch cold by going outside with wet hair, and the recent exponential rise in the reporting of food intolerances comes in large part from modern folklore. Grisi siknis is presented here not as an oddity, but as a reminder that the meaning a person applies to a symptom must be understood for the problem to be addressed, and that is no different in Western cultures than it is in Miskito communities.

I am obviously not suggesting that the solution to psychosomatic disorders lies in attributing them to a 'spiritual' cause that can be washed away with a ritual. But I am saying that Western cultures might benefit from looking at their socially constructed ways of responding to illness and explaining bodily changes, and asking if they are effective. If the medicalization of bodily changes is creating chronic conditions and a dependency on the pharmaceutical industry, that would be worth knowing, especially if greater insight into the cognitive mechanisms that produce illness was enough to change the course of that illness.

Talking to Madda and the people of Port Arthur, I found a great deal in grisi siknis that I could admire. It can be a very effective culturally agreed means of expressing distress. It is an acceptable way to exteriorize and deal with personal and social conflict. It is also a useful one, because it comes without blame. The demon infiltrator presents an external cause that removes the focus from the individual. It also provides something at which to aim treatment. People who rely on Western medical formulations to explain feelings of ill health also use this method when they encounter psychosomatic symptoms. Blaming somatic symptoms on a virus or food intolerance gives the same benefit as blaming them on the *duende*, although Western formulations risk the drawback of being incurable.

I also found myself admiring the Miskito's relaxed attitude to labelling and systems of illness classification, both of which are objects of obsession for many Western medical institutions. Seizures that have a psychosomatic cause have been subject to numerous name changes: hysteria, pseudoseizures, non-epileptic attacks, psychogenic non-epileptic seizures, dissociative seizures and functional seizures. 'Psychosomatic' has fallen out of favour in neurological circles, with 'functional' taking its place. The term 'functional neurological disorder' (FND) is used to imply that the brain is not functioning – therefore (rather ham-fistedly, I would say) placing the source of psychosomatic disorders firmly in the biology of the brain. In a sense, the term has been created to remind the sceptics that psychosomatic disorders are real neurological conditions. To that end, all things psychological and sociological have been removed from the name, leaving a descriptive label that is less alienating than its predecessors. However, I would argue that it also gives the impression that FND is a purely biological illness, which is surely just another sort of dualism. In fact, while biological changes are essential to produce the symptoms, behavioural and psychological factors are also essential triggers or driving forces in the development of disability.

People feel very strongly about which is the right term to use. Since there is no perfect label, I will continue to use both 'psychosomatic' and 'functional' as I write, and hope that the stories of real people will transcend the imperfections in each. For me, grisi siknis demonstrates that labels do not need to be everything. A potentially pejorative label ('crazy sickness') has been neutralized by the fact that the disorder is not deemed to be a personal one, and the name has gained acceptance within the community. A great deal of effort has been put into finding a more acceptable term with which to refer to biopsychosocial conditions, and I think Western medicine could learn from the Miskito on this point too. To truly destigmatize psychosomatic conditions, it is better to approach society's understanding of them – changing the name is just rebranding. We have no chance of improving the acceptability of the disorder if the disorder itself continues to be regarded as the fault of the individual.

I met the younger generation of Port Arthur's Miskito in Starbucks. Most were born in the USA or had moved there in infancy. They had all been to Nicaragua on family holidays, but they hadn't lived there for any substantial amount of time. Two of the men wore beaded necklaces and had ear jewellery that gave them a tribal appearance. One woman, Saria, wore a hijab. She and her US-born Miskito husband, Elasio, had recently converted to Islam.

I started by asking the group if they considered themselves to be predominantly North American, Nicaraguan or Miskito. Alfredo, the most beaded and tattooed of them all, said, 'It's all just borders, ma'am. I don't see that it matters. But, if I'm asked and I have to say, I tell people I'm Native American, from the Caribbean.'

Alfredo was born in Nicaragua and moved to the US when he was three.

'How often have you been to the Mosquito Coast?' I asked them.

I got a smattering of answers ranging from 'once' to 'several times.'

'And you've heard of grisi siknis?' I said.

'Yes, ma'am. In 2009, my little brother had it,' Elasio told me.

'Did he get it while he was in the USA?'

'No, we were visiting Bilwi.'

'Was he born there?'

'No, he was born here, like me. It was our first time there. It happened in a taxi. We had been there a week. He just suddenly started screaming. His hands turned in. His eyes looked white and they were twitching.'

Bilwi is a town in Nicaragua, and Elasio's family had been there on a month-long holiday. I asked if he or his brother had been familiar with the condition before his brother was affected.

'Yes, ma'am, we knew about it,' he said with a matter-of-fact tone. 'Our parents tried to hide traditional beliefs from us. In the USA, they don't want us to know about things like demons, but I've always been interested in where I came from, so I've read about it.'

I realized these young people were not typical of all second-generation Miskito in Port Arthur. They spoke Miskito at home, whereas most of their Miskito contemporaries did not. Alfredo and Elasio had read a great deal about the Miskito culture, including books dating back to the nineteenth century.

'Were you surprised when your brother got it?' I asked.

'Yeah. I was sceptical about it before I saw it. I grew up in a scientific community. I didn't believe in things like witchcraft. But, now I've seen it, I know it's true.'

'Why did he get it, do you think?'

'Because he was a stranger in Bilwi. The forest spirits are suspicious of outsiders.' Elasio's answers were all emphatic.

'He might have stepped on *pueson bikan* without knowing,' Alfredo suggested.

Pueson bikan, he told me, as he scribbled the words in my notebook, was buried poison. It could have been placed there to target an individual or because somebody was experimenting with witchcraft, and Elasio's brother became its unwitting victim. (I later asked Madda if she had heard of anything like this and she told me that *puisin*, as she spelled it, refers to sorcery artefacts in general and not poison in particular.) I asked them about the black book, which was a recurrent theme in the older generation's stories. They hadn't heard of it. They attributed grisi siknis to what is known in Nicaragua as a '*lasa*', which they told me could be either demon or deity in their culture, and I later read that lasas are also known to have been deities in the Etruscan culture of ancient Italy. Elasio explained that he believed lasas existed in a different realm to humans, but that they were as real as people. He called them 'unseen creatures in nature'. Lasas, he said, could be forest or water spirits, as well as demons.

I asked if there were other causes, and, without prompting, the group echoed the previous stories I'd heard, saying that grisi siknis could be caused by older men chasing younger girls. Some men made love potions, they said.

'Was your brother sick for long?' I asked Elasio.

'We took him to a shaman. He treated my brother with local flora and he got better quickly. It was gone in a day.'

'Do you talk to your US friends about the traditional Miskito beliefs?'

'Mostly no. I will if somebody seems interested and seems able to be respectful. As Charles Napier Bell said, they have knowledge from books, but they don't know anything about nature.'

Charles Napier Bell was Scottish born, but spent much of his childhood in the Mosquito Coast. In 1899, he wrote a book, *Tangweera*, about his life there. Elasio was clearly drawn to the Miskito culture and had read a great deal on the subject, but he had also converted to Islam. That seemed strange to me.

I asked him how he'd come to convert.

'I didn't like how Christians treated my people, ma'am. They destroyed indigenous culture. They destroyed our language. We lost a lot of words. They are hypocrites. They made fun of our beliefs, but at the same time they had no problem when burning bushes talked to people in their Bible. A lasa, to the Christians, was just negative, a demon, but, before the Christians, a lasa was a positive thing, too.'

Alfredo stepped in. 'The elders are colonialist Miskitos. They prefer Christianity. They are not natural Miskitos. They would be ashamed of the natural Miskito, if they were standing beside them. In the old days, the natives stretched their ears with eggs. The elders don't even know things like that. I feel closer to a natural Miskito.'

Alfredo had stretchers in his earlobes.

I turned to Amara, who had been sitting listening politely. A little older than the others, she was an accountant who was visiting from the Nicaraguan capital, Managua, and who was born on the Mosquito Coast. In contrast to her North American friends, who were bedecked in tribal beads, she was dressed more fashionably and would not have been out of place at a trendy London or New York gathering. She was the only person I had met who was still living in Nicaragua, and I had been waiting to hear her point of view.

'Grisi siknis is like a dream that cleans from the inside,' she said.

Beautiful, I thought.

'Have you had it?' I asked.

'Nearly,' she said.

When Amara was a teenager, her sister and cousin had grisi siknis. They had the characteristic super-strength and tendency to run away, so had to be locked in. The family all lived together, and neighbours advised them that Amara should move away.

'I was sure I wouldn't get it, so I refused to go,' she told me.

But she was almost proved wrong. She had felt it creep over her and had to use every ounce of her inner strength to ward it off.

'Do you worry you'll get it again?'

'No; I'm too old now.' She laughed.

One by one, the group told me of their beliefs. Spirits and demons and witchcraft sat alongside science. Elasio's sister has cerebral palsy and epilepsy. He and his family had learned the language of neurology to make sure she got the best care. He wanted to bring medications for epilepsy – as well as ayahuasca, the hallucinogenic herbal treatment – to Nicaragua, to test their effect on grisi siknis.

The different versions of grisi siknis that I heard described said something about the fluidity of culture. The stories of it felt fond, but also in flux. I knew that, if I spoke to people still resident in Nicaragua, I would hear something different again – a bleaker version, I guessed. The older generation of Miskito resident in Port Arthur were washed in the nostalgia of expatriates who still had a hankering to live in the jungle, with a view of the Caribbean. Elasio and his friends told a twenty-first-century story, in which they married their US upbringing with their ancestral history, which they had gathered carefully from wherever they could glean it. Psychosomatic symptoms have a social life that moves with the times – as these people, young and old, were doing. Grisi siknis gave young girls in Nicaragua a moment to step out of the conservative, constrained role they had been given. The young people of Port Arthur had entirely different concerns. Instead of asserting themselves with illness, they did so with body art and jewellery and religious expression. Elasio, Alfredo and their friends were intensely proud of their cultural heritage, but their rebellion, if one looked closely, was a North American one.

Human patterns of behaviour follow the routes that are available to them. Those back in the villages of the Mosquito Coast were experiencing the reality of what was being described, and doing so in living circumstances that were significantly more deprived than the lives of their emigrant families. Most Miskito villages are small and have limited, if any, modern healthcare. But a church, a pastor and a shaman are always available. A person looking for help must keep in mind what help there is.

3

Paradise Lost

***Expectation:** The degree of probability that something will occur.*

Some years before I met her, I had seen Lyubov pictured in an online news article. She was a small, plump woman in late middle-age, with maroon-dyed hair and pale-blue eyes. The photograph showed her seated on her bed, wearing a brightly patterned housecoat. Hanging behind her on the wall was a green dress adorned with huge purple flowers, and even her bedspread had an ornate floral design. In contrast to the flamboyantly coloured scene around her, Lyubov wore a solemn expression as she stared away from the camera and into the distance. I learned from the accompanying article that she had been in and out of hospital eight times in five years, with a mystery illness. She was not alone: 130 of her neighbours had also been affected. All of the victims lived in one of two small towns in Kazakhstan, called Krasnogorsk and Kalachi.

Lyubov's story had come to mind when I read about Sweden's sleeping children. Although thousands of miles apart and with wildly different lives, the two groups seemed to have more than a little in common. Lyubov had also inexplicably fallen asleep, albeit for a much shorter time than Nola or Helan. I emailed the journalist who had written the article about her, and a year later I was on a plane travelling towards an abandoned town in the middle of the Kazakhstani steppe.

Krasnogorsk and Kalachi lie 300 miles north-west of Nur Sultan, the capital of Kazakhstan. I flew into Nur Sultan on 8 June 2019, on the eve of the country's first democratic election since it gained independence from the Soviet Union in 1991. It was an interesting time to find myself introduced to a place that was going through such change, and about which I knew so little. In March 2019, Nursultan Nazarbayev, Kazakhstan's only president since independence, had resigned. He had been appointed by the Soviets, and he in turn appointed his successor, Kassym-Jomart Tokayev. The elections promised to be the Kazakh people's first opportunity to choose their own leader; a chance to validate Nazarbayev's appointment of Tokayev – or not, if that was their preference.

Nur Sultan has only been Kazakhstan's capital since 1997. Until then, the capital was Almaty, a cosmopolitan sprawl in the south. A green city that sits snuggly against snow-capped mountains, it is dotted with parks and tree-lined avenues, making it a pleasant place to live. Almaty ceased to be the capital when, by presidential decree, Nazarbayev bestowed that honour to a place called Akmola, 750 miles north, in the geographic centre of the landlocked country. Until it became the capital, Akmola, which means 'white grave', was no more than a remote outpost of the former Soviet empire, most noteworthy for being the site of one of Stalin's gulags. It was surrounded by hundreds of miles of underpopulated steppe, icy cold in the winter and mosquito infested in the few short weeks of summer. The move from the verdant south to the brutal north was sold as an opportunity to attract investment to an underdeveloped area of the huge country. Cynics would say that it was more likely the desire to attract Kazakhs to the north to dilute the majority Russian population who lived in the region.

When it became the capital, Akmola was renamed Astana (rather unimaginatively, as it simply means 'capital' in Kazakh) and a phase of rapid building began. In the two decades since it was founded, it has been something of an architect's playground, and has evolved into a city with one of the most surreal skylines in the world. Buildings take the shape of pyramids, spires, spheres and tents, with as many facades as possible being clad in gold. In keeping with the otherworldliness of the city, the nickname given to the globe-shaped Nur Alem Pavilion, where many international conferences take place, is the Death Star.

The decision to rename the capital for a second time, from Astana to Nur Sultan, was made in March 2019 by Tokayev, in honour of the man who had chosen him as his successor. Three months later, on my arrival from London in June 2019, the name Astana had already been erased from all buildings and signs and official documents, as if things had never been any different. I was nervous and excited to arrive just in time to see Kazakhstan carry itself into democracy, or at least that's what I thought was going to happen. I had travelled from a London which, at the time, was overrun with protest and visible dissent. I expected to see the same in Nur Sultan.

On election day, I went into the city centre to get a sense of the atmosphere. I found it eerily quiet. The polling stations were open and the shops were closed. People wandered around as if there was nothing special about the day. There were none of the placards or loudspeakers that mark election time in the US or UK – there was no excitement of any kind. I eventually found a central square where people had congregated. Even there, the scene was orderly, without any clear indication of the purpose of the gathering. It was only after I had watched for a while that I saw much of the crowd was being directed, one by one, onto a bus, like an organized tour, only they weren't being shepherded aboard by tour guides, but by the police.

The biggest culture shock for me in a new place comes through the stifling effect of not speaking the language. In Kazakhstan, I couldn't even read street names because I didn't understand the Cyrillic script, so newspapers were impenetrable. In any new place, cultural subtext is hard to decipher, but in Nur Sultan I couldn't even scratch the surface of how the people were feeling about this politically charged day. Mobile Internet has made the world very small, so I turned to that for news of the election, but the pages for popular English-language news websites all froze when I tried to open them. I turned to social media, but those sites were also inaccessible. It took several minutes for me to accept that the websites I was trying to access had all been blocked. At last, I realized that I was having my first experience of a censored Internet, and it dawned on me that people can't congregate if they can't communicate.

The shutdown only lasted twenty-four hours. The next morning, social and news media was up and running again, but not until the flashpoint of potential democratic demonstration had passed and the election had been won by Nazarbayev's handpicked successor, Tokayev.

The thirty years that followed the dissolution of the Soviet Union, that period that had changed the 'white grave' into Nur Sultan, had had the opposite effect on Krasnogorsk. The politics of the country had been fundamental in creating for the people of Krasnogorsk the circumstances that would lead to the town's downfall. In the 1960s, approximately 6,500 people lived in Krasnogorsk. In 2010, the population was closer to 300.

By 2019, only thirty residents remained. The slow death of this small town, and that of its neighbour, Kalachi, is best understood by listening to the personal accounts of the people who used to live there, as they tell the tragic story of the mystery illness that drove them from their homes, turning Krasnogorsk and Kalachi almost into ghost towns.

It was with the knowledge that I was in a country very different from my own that I set off to meet Dinara, a local journalist who, through a series of emails, had agreed to arrange meetings with the people of Krasnogorsk and to be my companion on the road. I'd spent several days in Kazakhstan as a tourist before our planned trip to Krasnogorsk, and had spoken to Dinara only once during that time, when I phoned her in a panic, having been thrown off a train for having an invalid ticket. As an official walked away with my passport, I quickly dialled the phone number she had given me in case of emergency and was very relieved when an English-speaking person answered. She rescued me, and my passport, through a brief telephone exchange with the train guard. Still, that being the sum of our contact, I was relieved again when I got to the meeting place at the train station and found her waiting exactly where she said she would be.

Like the majority of people in Kazakhstan, Dinara is ethnically Kazakh, a group descended from Turkic and Mongol tribes. They make up 67 per cent of the population of Kazakhstan, with people of Russian descent coming second, at 20 per cent. Dinara was tall, with a broad smile that immediately put me at ease. We got to know each other sitting in the buffet car of the train that took us to Esil, the closest sizeable town to Krasnogorsk and Kalachi. Having spent a week alone, getting lost and ordering from menus I couldn't read, it felt good to be under Dinara's authoritative wing. During our four-hour journey, we gave each other our potted life stories.

It had taken several weeks of cold calling for Dinara to secure interviews with people from the sleeping towns. As we chatted easily together, and as I watched her do the same with every person who passed our seats, I saw the confident, friendly manner that I'm sure was crucial to her success in arranging meetings with people who had grown wary of journalists, having been besieged during the height of the illness. Dinara lived in cosmopolitan Almaty, a thousand miles from our destination, and I learned that Krasnogorsk and Kalachi were almost as strange a prospect to her as they were to me. Kazakhstan is a big country, and where we were going was not the sort of place one visited for fun.

Most of our meetings were to take place in Esil. When the train finally pulled in, I found myself a world away from where my journey had started, in Nur Sultan. Esil was tired, grey, flat and unattractive. Dinara and I hopped from the train and she flagged down a taxi. Our first appointment was with Tamara, who had been relocated from Krasnogorsk to Esil when she fell ill.

The taxi was an old orange Lada that I would never have recognized as a taxi had Dinara not been with me. It had torn seats and the inside door handle was hanging off. Dinara sat in the front and instructed the driver, and I sat in the back, clinging nervously to the grab handle to make up for the lack of seat belts. Watching Esil through the window, I thought it seemed a functional place. Lots of concrete, built for purpose and not for leisure. It was midday, but the streets were almost deserted.

Tamara's apartment block was nondescript and grey, like the

town. We took the narrow concrete stairs to the second floor and knocked. When she opened the door and I saw her for the first time, I was completely taken aback. She did not fit at all with the austere, superficial impression I had formed of the place where she lived. She was a seventy-year-old woman, yet she had the obvious glamour of a fading movie star. Her long, wavy blonde hair was swept to one side, so it sat mostly on her left shoulder. She wore gold-rimmed glasses, gold teardrop earrings and a black and white zebra-print dress. She was carefully made up with bright-red lipstick and red nail polish, on a cold, windy Tuesday afternoon.

Her apartment was special too – very small, but full of reminders of the past. The walls were covered in photographs – many of her family, and several of her when she was younger. Her daughters were also very blonde, while the men and boys in the family were darker. In nearly every picture, the family's expressions were solemn, like in old black and white photographs, before smiling for the camera had come into fashion. Tamara's hair always sat artfully on one shoulder, as it still did. Her strong bone structure meant that, at seventy, she had changed very little since those younger days.

The sitting room was decorated in brown tones, there was a paisley carpet hanging on the wall, a faun and cream velour sofa and a cabinet in which ornate crockery was displayed. Several oversized soft toy animals sat waiting for Tamara's granddaughter, who lived in Russia, to come and visit.

'The decor,' Dinara said quietly to me, 'it's very Russian.'

Perhaps it is, I thought, although, when I was growing up in Dublin, we also had a velour sofa and a sideboard with rarely used crystal glasses and china plates on show.

I settled myself on the sofa beside a large pink stuffed elephant. Tamara sat opposite, straight-backed, telling her story to Dinara, who, in turn, told it to me. It began on 1 October 2015, when Tamara had fallen asleep and couldn't be woken for forty-eight hours.

The first sign that something was wrong happened at a party in the cultural centre in Krasnogorsk, Tamara told me. She had lived in Krasnogorsk for nearly fifty years – the happiest years of her life were spent there. It was her beloved home, where she had raised her children, but she is now one of its many refugees. The party was a community affair and most of the town residents were present. Halfway through the evening, to her surprise, Tamara started to feel strange. Her head felt light and she was unusually sleepy, although it wasn't late. She didn't think much of it at the time, although it was bad enough for her to have to leave the party early. When she got home, she looked in the mirror and thought she looked tired. She went to bed, expecting to feel better the next day. Instead, things were significantly worse. When her husband got up for work, he couldn't wake her. She didn't look especially unwell, she just looked asleep. It took several minutes for him to accept that it was more than an ordinary lie-in, at which point he called the doctor. Tamara did rouse briefly while they were waiting for the doctor, but she couldn't stay awake. She was back in bed, fast asleep again, when the doctor examined her. He couldn't find an explanation for her condition, so he called an ambulance to take her to the local hospital. When it arrived, Tamara did something very strange: she appeared to wake, then got out of bed, went to the mirror and fixed her hair and make-up. She then returned to bed. Much later, when she had finally recovered, her family would laugh about it.

'Even sick, I won't go out without my make-up!' she told me. Until this point in the story, she had looked sad, but this made her laugh.

'And you don't remember doing that?'

'No. I woke up in hospital. My husband said it was as if I was operating on automatic.'

Tamara was in hospital for two days. She remembered little of it. Her family told her that she had been asleep for the majority of the time. Doctors did a range of tests, but couldn't find anything wrong. When she finally awoke fully, her head was spinning and she couldn't stop hiccupping. She was very unsteady on her feet and it was another day before she could walk properly.

'What happened in hospital, when you were unconscious?' I asked. 'Could you eat? Did you go to the toilet, or just lie in bed?'

'I don't know. Most of my time in hospital is a blank,' she said, 'except when the deputy mayor came to visit me.'

She looked very pleased when she told me about her honoured hospital visitor. Tamara had some status in Krasnogorsk: she had worked in its cultural centre for decades and counted the deputy mayor among her friends. When he had heard she was unwell, he rushed to see her. He ensured that she was getting the best care, and she had woken briefly during his visit.

The worst of Tamara's symptoms passed fairly quickly, once she was awake. Within five days, she was well enough to go back to work, although she still didn't feel fully recovered. In fact, she never went back to feeling as healthy as she had been before she fell asleep.

'Look at my hands,' she said to me. 'They're cracked. They used to be beautiful.'

She gave off a forlorn air of loss. She was surrounded by reminders of her youth and mementos of her family, but she lived alone and it felt sad. I wondered how that had contributed to the sleeping sickness. But if it was sadness converted into a need to sleep, much as Freud might have suggested, how did it spread from person to person?

'Had you seen other people with the illness before you got it?' I knew Tamara was not its first victim.

She told me she had seen a young girl collapse some weeks before the night of the party. She had also heard the stories of the sickness circulating the town, although she hadn't paid them much attention.

Interpreting a medical history through a translator is a frustrating process for a doctor. The nuance of the symptoms is lost, and I was struggling to understand exactly what had happened to Tamara. In psychosomatic illness, the devil lies in the small details, in the turn of a phrase. This sleeping sickness, like resignation syndrome, was also geographically contained, affecting only two neighbouring towns, Krasnogorsk and Kalachi. Culture-bound syndromes are often a metaphor for something that cannot be expressed in a more explicit way within a certain community. Grisi siknis allows girls to express themselves in a society that imposes conflicting values on them. Resignation syndrome gives a voice to the voiceless. If the sleeping sickness was psychosomatic, what was it about these two small towns that had created it? I was trying to appreciate Tamara's experience, but could only hear Dinara's voice. Sometimes Tamara spoke so quickly and for so long that Dinara had to make rapid notes, which she summarized for me minutes later. Frustrated, I asked Dinara to tell me precisely what words Tamara had used to describe how the illness felt. Dinara asked her to describe it again.

'She said it is like a person becomes a trained reflection of themselves. The body can be awake, but the brain is not, so the person's understanding of the world is paralysed.'

Tamara had done her make-up and fixed her hair before she went to hospital – in that moment, she was an echo of herself.

'Did the doctors give you any specific diagnosis or explanation?' I asked.

She went to a cupboard and pulled out a pile of paper that had been tucked into an official-looking envelope.

'These are her medical records,' Dinara told me. 'Everybody has a packet like this that they keep at home.'

'Do you have one?' I asked.

'No!' She laughed. 'I've never been great at keeping records!'

I leafed through some of the papers. Although most of it was in Russian, medical words were in Latin, so I could recognize them without help. The diagnosis was encephalopathy, a fairly generic term for any type of confusion. It was more a description of her clinical state than an actual diagnosis. Dinara translated the rest: *skin and tongue, dry; abdomen, normal; heart and lungs, normal; lymph nodes, normal; blood tests, normal; CT scan of the brain, normal; toxicology screen, normal. Cause: unexplained.*

'Was there much alcohol at the party?' I asked.

Typical doctor question. I simply couldn't help myself.

Tamara laughed. 'No – it was a party for old people. Much older than me. It was more like a house party or a picnic. People drank tea. Everybody brought salads and cakes to share.' She thought for a while. 'I shouldn't have been surprised to get sick, that night.'

'Why not?'

'The sleeping sickness always happens in places where there are lots of people gathered together,' she said.

The problem started in 2010. Lyubov, whose picture I had seen in the news article, was the first victim, I learned. Nadezhda, a nurse, was second. After them, the outbreaks had come in waves, with every case as mysteriously unexplained as Tamara's had been.

'They took samples of my hair and my nails, and sent them away. They said they found nothing,' she told me.

Between 2010 and 2015, approximately 130 people, from a population of just over 300, had fallen ill with symptoms like Tamara's. She was one of the last victims. Shortly after her recovery, the mystery illness disappeared as abruptly as it had appeared. When I read about the disorder in newspapers, it was usually referred to as a sleeping sickness, but, talking to Tamara, I learned that the symptoms were much more complex than that. Certainly, the most common feature was sleep, particularly in the older people affected, but there was a variety of other presentations. It was very different in children, for example. Rather than sleeping, they behaved strangely and many laughed uncontrollably. Cases emerged in clusters – in one instance, nine children in a single class were affected on the same day. Far from being subdued, they ran around manically, they hallucinated, they fell to the ground in convulsions. I asked about the content of the hallucinations, but Tamara didn't know. She hadn't experienced them.

It seemed the disorder had a very wide range of forms. Some people lost the ability to walk or talk, but didn't sleep. Acting on automatic was a common manifestation. A person could appear to be asleep, but then they would wake, suddenly answer a question entirely appropriately, and then fall straight back to sleep again. The illness affected men, women and children. It could last a day or several weeks. Unlucky people had recurrent bouts.

All of those affected were investigated at the local hospital. When the doctors there couldn't find a cause, some of the sicker people were sent to Nur Sultan (then still called Astana) for more sophisticated tests. But those tests also came back as normal. A handful of the victims were sent to a hospital in Russia, but the outcome was the same: no cause found. The good thing was that everybody affected eventually woke up spontaneously. They were given various diagnoses, with toxic encephalopathy being the most common one. But, if there was a toxin, nobody knew what it was.

Once the local doctors had raised an alert, the Kazakhstani government became involved. Krasnogorsk had been a mining town. Experts came to test the soil and water for contaminants. The mine had been closed for years, but still they checked the air inside the mine. All the tests came back clear. Representatives of the Institute of Radiology spent weeks in Krasnogorsk investigating, but drew a blank.

The sickness continued to spread systematically through the town and it became a national talking point. President Nazarbayev made a personal appearance on television, calling for foreign researchers to come to Krasnogorsk to explain the mystery. I don't know if any researchers answered his call, but foreign journalists certainly did. The town was inundated and its inhabitants were interrogated. International newspapers and websites ran pictures of the locals in front of dilapidated buildings, staring wistfully into the distance. Throughout this period, the residents of Krasnogorsk continued to fall ill – but none of the visitors did. Even the government researchers who spent a very long time in the town remained symptom free. Like resignation syndrome, the victims of this outbreak were highly selected.

'Did animals get sick?' I asked.

'No.'

'What do you think caused it?'

'I think we were being poisoned,' Tamara told me firmly.

'Poison from the mine?' I asked.

'No, the government.'

Her answer surprised me. 'Do you mean deliberately poisoned?'

'I think the government poisoned us because they wanted us out of the town.'

I didn't know what to say to that. I always believe accident and incompetence to be more likely than conspiracy, so I asked again, 'Why not from the mine?'

'That had always been there. It never made us sick before.'

'But why would the government want you to leave?'

'I don't know.'

'If that's true, what do you think they poisoned? The air? The water? The food?'

Tamara mulled over the question. 'It couldn't be the water. Everybody drank the water, but not everybody got sick.' She thought for a while more. 'But it can't be in the air either.'

She didn't know what had been poisoned, but she was sure that poisoning was the cause. On this point, her sadness was replaced by bitterness. She was furious at the injustice.

In my head, I kept coming back to the why of it all. Clearly, this was a country whose politics were alien to my own experience of the world. I could not and would not dismiss her fears about the actions of the government. But Krasnogorsk was a tiny town, hundreds of miles from any major city. I couldn't make sense of the conspiracy.

'What sort of place is Krasnogorsk?' I asked. I hadn't yet been there, though I was due to visit.

At my question, Tamara relaxed her posture, took a breath and smiled. 'Paradise,' she said. 'Krasnogorsk was paradise.'

I was a little shocked. I considered Esil drab, on the surface at least, but, from the pictures I'd seen in several news articles, Krasnogorsk looked even worse.

'Did she use the actual word, "paradise"?' I checked with Dinara. She nodded, yes.

'We were so happy there,' Tamara added. 'It was such a special place. But this place . . .' She pointed out of the window at the grey deserted streets of Esil, shook her head and frowned.

Tamara was born in Siberia. Paradise, I suppose, is relative. Krasnogorsk was a mining town, and her husband had been sent there to work as a mining specialist. She was in her early twenties, with a seven-month-old son, when they left Russia and moved to Kazakhstan. She was a trained dancer, she told me, and had attended a Russian institute of choreography. It had not been her choice to move to Krasnogorsk, but she was relieved, when she got there, to be given a job in the cultural centre, teaching dance. Eventually, she became the director of cultural events.

She took her laptop from a desk and sat it in front of me, her face coming alive. 'Look at my girls,' she said, and showed me a video of teenagers dancing – first ballet, and then something that looked like belly dancing, performed in high-heeled shoes. Tamara had been the girls' teacher. Showing me the videos transformed her again, this time into somebody proud and happy.

'These are some of my girls. It was such a wonderful life, there.'

During Soviet times, Krasnogorsk had several thousand residents. By the time the sleeping sickness hit, there were fewer than 300. If it had ever truly been a paradise, I struggled to believe it still was when Tamara fell ill, in 2015. It must have felt so empty, all those buildings and so few people. I hinted at this to Tamara, but it only provoked her to repeat that Krasnogorsk was a green and perfect place, and that being forced to leave was the worst thing that ever happened to her.

'The water here isn't good. It tastes wrong,' she said of her new home.

Not like the water in Krasnogorsk, which – like the trees, the garden, the river, the people, the homes – was exactly right. In Tamara's memory, at least.

I knew from newspaper reports that the mine which sustained the Krasnogorsk economy had been shut down in the 1990s. Almost the whole town became unemployed overnight. People left to get jobs elsewhere – not that there were any jobs available. This was immediately after independence from Russia, and Kazakhstan had long ago forgotten how to look after itself. The closure of the mine had been almost immediately followed by the creeping loss of normal amenities in Krasnogorsk, and, despite Tamara's glowing account, I knew that most of the town had had no running water for years. Many homes had no heating, in a place where the temperature dropped to minus fifty degrees in the winter.

But, as Tamara made clear, these hardships were nothing to her and she had never wanted to leave. The sleeping sickness had forced her hand. When the epidemic took hold, the Kazakh government offered relocation to the straggling residents of Krasnogorsk. Apartments were made available in Esil and people were encouraged to move. Tamara refused. They all refused. It was only when Tamara got the sleeping sickness for a second time, and her husband and son caught it too, that she agreed to swap her three-bedroom apartment in Krasnogorsk for a much smaller one-bedroom apartment in Esil.

'She's angry,' Dinara told me, when Tamara went to make us tea.

'I'll be honest,' I said, 'I don't get why she liked the place so much. It sounds kind of in the middle of nowhere and pretty bloody bleak, to me. They had no running water! I would have been longing to leave.'

'This apartment is half the size of her old one,' Dinara countered. 'No one wants to go down a rung.'

'But this flat has a radiator and a toilet that flushes.'

'She's absolutely convinced there was a conspiracy to clear the town,' Dinara advised me.

'But why would anyone do that?'

When Tamara came back with black tea in china cups, Dinara put the question to her again. She didn't know why, she told us. She was mystified too, but still she was certain that it was true. She went to the sideboard, pulled a letter from one of the drawers and showed it to me. It was a list of names, typed out on official-looking headed paper. A record of all those affected by the sleeping sickness. One hundred and thirty-three names, although she believed there were many more who had not made it onto the list. They planned to sue the government.

Maybe that's what this was, I thought: a way to get a better exchange for their homes. My mind was still lingering on the word 'paradise,' though. I asked if I could see a photograph of Krasnogorsk from before the mine closed. I had seen many recent pictures in the newspaper, but none from the time when the town was thriving. We looked through her photograph album, but all the pictures were of people, never the place.

'Can I take a photograph of you?' I asked Tamara, as the conversation was drawing to a close. 'Not to publish, just to remember you better.'

She refused. Instead, she asked that I take a picture of one of the photographs of her when she was young.

'We all have a time when we look our best,' she said. 'I am getting old now.'

I moved around the room, taking photographs of photographs.

I finished by asking, as carefully as possible, whether she thought it possible that there could have been a psychological cause for the problem. Perhaps life in a deprived, dying town was just too hard?

'*Niet, niet, niet*,' she said.

I didn't need Dinara's help to translate that.

'It must have been a very difficult life?' I said, never knowing when to give up.

'There was not a difficult life. Here, in this dusty town where I don't want to be, is the difficult life.'

After we left, Dinara and I pored over what we had been told. Poison seemed very unlikely given that people had had such varied symptoms and nothing showed up on tests. While a poison might be undetectable, its effects should have been seen in abnormal blood tests or as brain-scan abnormalities. When a person is unconscious, their brain should look unconscious, even if the cause is a mystery. In other words, if a person has been poisoned, the end result of that should be detectable and there should be some correlate to the symptoms on medical investigations, but there was none. Extensive investigation of the residents of Krasnogorsk showed no objective evidence of a disease process, no matter how sick they were. No animals got sick. No reporters. No scientists. No government officials. Recovery was spontaneous on every occasion, even for those few people who stayed in the town.

Nor did the disorder make any sense in terms of pathology. Disease reveals itself, even if the cause can't be determined, and this outbreak defied scientific reason. The majority of waves of illness happened at group gatherings, but not in any particular part of town. Not always indoors or always outdoors. The occasions were all different, some involving food, others not, and at any event only a handful of people were affected, even if they had all eaten the same food, drunk the same water and breathed the same air.

Five years had only seen an evolution in symptoms. This is absolutely typical of a psychosomatic disorder: it evolves over time as its story is retold by each new person affected. I had seen that evolution in the stories of grisi siknis, where each generation added new elements and interpretations. In the case of Krasnogorsk's sleeping sickness, different people experienced it differently, and in children it was wildly different from how it manifested in adults, almost as if they had another problem entirely. Doctors' and scientists' investigations into the matter had been fairly exhaustive, yet they found no objective proof of a toxin or a virus, and none of disease. Of course, Tamara might argue that the investigations were carried out by the government, and she didn't trust them. I had already learned during my brief stay in Nur Sultan that there was some justification in her mistrust. Still, it seemed the government hadn't used Krasnogorsk for anything after the people left, so it was hard to see what they had gained, if the accusations were true. A psychosomatic condition seemed inescapable, to me. The symptoms and disabilities made no anatomical or physiological sense. The outbreak was inconsistent with disease. I thought of the convulsing and hallucinating schoolchildren that Tamara had mentioned. In Nicaragua, they would have been told that they had grisi siknis, but, in ex-Soviet Kazakhstan, spiritual explanations didn't exist, whereas toxic mines and devious governments did.

Reading about the outbreak before travelling to Kazakhstan, I had felt that the harsh life the people lived must have been at fault. The people were unemployed and living in extreme poverty. There had also been a large influx of journalists during the peak of the outbreak and it seemed to me that their arrival would have enlivened the town again. The stress of everyday life had caused the illness and the arrival of the media could have perpetuated it. Tamara's story, and her absolute denial that hardship was at the core of the condition, had done nothing to change my mind.

As we waved goodbye to Tamara at the door of her apartment block, Dinara took a phone call.

'A doctor has agreed to talk to you,' she told me, when she came off the phone. 'He treated some of the people.'

I was pleased; I wanted a more objective viewpoint, although I was aware the doctor had been reluctant to give an interview. It had taken all of Dinara's powers of persuasion to get him to agree, and he asked not to be named.

The hospital, like all the concrete buildings in Esil, was so lacking in personality that I would not have known its purpose if I hadn't been told. The only clue was an exceptionally old, grey, Soviet-era ambulance parked outside – a vehicle that looked more like a camper van. The doctor, a small man in his forties, met us at a fire-escape door at the rear of the building. It felt covert, but I think it was just a convenient meeting place because his office was nearby. We shook hands, and I could see by his expression that he was reticent. He led us to his office.

'What is your book about?' he asked, as soon as we were seated.

After I had explained, he still looked suspicious. A lot of reporters had visited the region during the time of the sleeping sickness, but, despite this, the people felt their message and their priorities had never been properly represented. They wanted to understand the government's motivation for getting them out and they wanted proper compensation for the homes they had lost.

'I am very busy today; I don't have much time,' the doctor said. 'How can I help you?'

I wanted to understand the typical symptoms of the disorder, so I started by asking him about those. He told me he had been involved in the care of several of the sleeping people. As he described their condition I saw for the second time that there was no such thing as a typical set of symptoms. People became sick in lots of different ways – each in their own way, one might argue. Even 'sleeping sickness' wasn't a very accurate label; some slept, but lots didn't.

'Children tended to be hyperactive,' he told me, corroborating Tamara's story. Sometimes they screamed, or appeared to pluck at the air as if trying to catch something that was invisible to others. They were also prone to vomiting and hallucinations.

'What sort of hallucinations?' I asked, thinking again about grisi siknis.

'One boy saw snakes; one girl thought she had worms in her tummy. A lot of people said they could smell something sweet.'

The people who were sleepy tended to sleep between two and four in the afternoon, he told me. The degree of sleepiness was very variable: some could be woken, but not for long; some snored very deeply and could not be roused at all. Fainting was common – although he didn't consider it to be 'normal' fainting; the affected person crumpled to the floor while retaining some control. Adults could behave strangely, but in a different way from the children. Some tried to climb out of windows, others walked around in their underwear. Their coordination was poor and many were off balance. The sickness attacked the town in waves. Some days, he admitted five or six people within a few hours. He agreed that it was particularly likely to affect groups attending social gatherings.

A large number of cases happened during a celebration for the first day of the new school year. He had examined one of the children. The child's parents had been called to collect her from school because she felt ill. During the car journey home she had fallen into an unnatural sleep. She stayed like that for hours. Similar to other victims, she would sometimes open her eyes briefly, but never for very long. Ultimately, she had been taken to the hospital where she finally awoke, but even then she had not immediately returned to normal. She was in a confused state and her behaviour was odd. She kept insisting on trying to help her mother, who was happily seated, to stand up. Her mother had neither asked for nor wanted help. The child was fully recovered by the following day but couldn't recall anything that had happened. All her medical investigations had been normal so her illness was never explained.

'What did the other people's tests show?' I asked.

When the first case had come in, he had thought it was caused by a stroke, but the patient's brain scan was normal. A couple had high blood pressure, while a couple more had raised amylase levels. Elevated levels of this digestive enzyme can indicate pancreatic disease but would not account for the sleeping sickness. There was no common abnormality. For the most part, all tests on blood, hair and nails were normal. They had checked for heavy-metals poisoning, but no trace was found. A test for carbon monoxide was positive in one person, he thought, but only one. He believed that some of the brain scans done in Astana had shown brain swelling, although he hadn't seen those results himself. When the people were admitted to hospital, they were given intravenous fluids until they woke up. That was the only treatment. Everybody affected got significantly better, but at different rates. Some had recurrent bouts. Quite a few, like Tamara, never felt 100 per cent again.

'What do you think caused it?' I asked him.

'Poison,' he said, with absolute certainty.

'But why was it so patchy? Where was the poison?'

'There were patterns,' he said. 'A large number of people in one street in Kalachi were affected. And an outbreak was more likely if the wind was blowing from Kalachi towards Krasnogorsk. Something from the mine?'

Kalachi and Krasnogorsk were very close to one another. Kalachi was a farming town, not a mining town, but one of the mines was located there.

'But scientists came and checked for toxins?'

The concern about the mines was not a silly one. These were uranium mines, albeit closed for many years. That was why I had been surprised Tamara did not implicate them. The government had been concerned enough to send a team of investigators to Krasnogorsk and Kalachi. A non-functioning mine should not be dangerous, but still the Institute of Radiology had been dispatched. Their officials spent a year in the town, carrying out air, soil and water tests. None of those investigators got sick, and neither were there any new cases of sleeping sickness recorded in the residents during the year the investigators were there. The doctor, like Tamara, expressed the concern that the government was deliberately poisoning the villagers to get them to leave.

'But why?' I asked again.

'It stopped when people from Krasnogorsk and Kalachi agreed to resettle,' he told me.

'But thirty people still live there,' I pointed out. 'Why stop poisoning people, if they haven't achieved the goal of emptying the town?'

He didn't know and seemed stumped by this question. I felt like he was deliberately ignoring inconvenient truths.

Most of those resident in Krasnogorsk in 2010 had eventually agreed to be moved to Esil. A few were still holding out for a better deal. Travelling to Esil by train, I had seen a vast expanse of unpopulated land. This was not a country that lacked space.

If the government had a secret development planned, they had room for it elsewhere. If they wanted to reopen the uranium mines, surely they would employ the old workers, rather than drive them out. Besides which, even after the people had left and the sleeping sickness had stopped, there was still no sign of the government's secret plan unfolding.

'Could some of this have had a psychological cause?' I asked tentatively, knowing how badly this question can be received. 'Could it have been a domino effect, where one person gets sick for some other reason, maybe even poison, which creates anxiety and it then snowballs?'

'*Niet*,' he replied.

Tamara and the doctor had answered most of my questions fairly willingly, but in reply to this question they had been equally brief and emphatic. I was mindful that the doctor had seemed reluctant to meet me and had already answered this question very clearly, but still I wanted to press him on the point. I explained my reasoning for preferring a psychosomatic explanation: the general impossibility of a poison that left no objective trace; the normal test results in unconscious people; the way the disorder picked off victims so randomly in crowds, but only affected people who lived in the town. A whimsical poison.

'My father, he was eighty-four, he was healthy all his life, but then he died suddenly. He should not have died,' he said, adding, 'I still think it could be the mine.'

The people of Krasnogorsk had very good reason to be concerned about living so close to a uranium mine. It would concern me. The World Health Organization has reported that high levels of radon are often detectable in uranium-mining areas. Radon is genotoxic, leading to cancer and DNA mutations. When the mine was open, the workers could certainly have been at risk from radiation. But the people had lived in the glow of the uranium mines for fifty years without complaint. In 2010, the mine had been closed for well over ten years. Why get sick then? Also, the type of sickness was wrong. Radiation could lead to lung cancer and birth defects, but it wouldn't explain the sporadic relapsing, remitting symptoms and sleep attacks. Members of the World Health Organization had been among the scientists to visit Krasnogorsk and Kalachi. They had also given the town the all clear.

An alternative explanation to radon was carbon monoxide poisoning, which certainly causes drowsiness, dizziness and unsteadiness, and could lead to coma. Chronic low-level poisoning would also be hard to detect. But once in hospital, away from the source of the carbon monoxide, and treated with oxygen, the sufferers should have woken fairly quickly, unless they had significant brain damage, which they did not, as was proven by the normal scans and EEGs. Even when in hospital, some slept for days or weeks. What's more, carbon monoxide collects in poorly ventilated spaces. It affects miners inside the narrow confines of working mines, not outside in the open air. And there would be no reason for carbon monoxide poison to have an increased chance of poisoning people at group events, which was a big feature of the sleeping sickness. Nor would it explain some of the odder features of the illness – acting on automatic pilot, crying uncontrollably, thrashing limbs. Also, where did the carbon monoxide suddenly come from in 2010, and where did it disappear to in 2015? Fear of a uranium mine was very legitimate, but that avenue had been investigated and thoroughly ruled out. Suspicion of a government that controls social media and serves its own purpose was also entirely reasonable, but still the conspiracy theory didn't hang together.

'Who was the first person to get the sleeping sickness?' I asked.

I had already been told it was Lyubov, but it was such an

important point that I wanted to be certain. In the case of mass outbreaks of psychosomatic symptoms, the first person to fall ill is often distinct from the rest. They get sick for their own personal reason, possibly with a totally different medical problem from those that follow. They could have a disease or a psychosomatic disorder; either way, they are the unwitting catalyst for everything that happens next. Their story is key.

'Lyubov was first,' the doctor confirmed. 'I thought she'd had a stroke, but it wasn't typical.'

'Did you like living in Krasnogorsk?' I asked. Throughout our conversation, I had the sense that he was uncomfortable. He was busy, I assumed, but there was something else. Perhaps he was angry at losing his home. He brightened momentarily when I asked him about it.

'I was happy there. It was a wonderful place. The hospital was well stocked and we had every type of specialist. Quality of life was good.'

Paradise, I thought to myself, remembering what Tamara had told me.

As we said goodbye, back on the fire escape again, the doctor gave me one final comment: 'All I want is compensation and follow-up care for the people who got sick,' he said.

Once alone outside the hospital, Dinara and I looked at each other.

'I don't get it yet, do you?' I asked her.

'No.'

'Actually, I don't understand most of the story. Why would a town with only 6,000 people have a hospital as good as the one he described?'

'Because of the mine, I suppose.'

It made sense – although, in such a poor and isolated region of the country, it also did not.

'I'm really excited to see the paradise that is Krasnogorsk,' I said as we waited for a taxi.

'Me, too. When I told my friends where we were going, they looked at it on a map. They said, "Why in the world would you go there?" Now, I can't wait.'

Another terrifyingly decrepit-looking Lada pulled up and we climbed in. Dinara began tapping away on her phone, trying to find new clues about the sleeping sickness in the local press.

'Look,' she said, showing me her phone. It displayed a newspaper report about a cat being affected.

'One sick cat does not a uranium mining incident make.'

'Cynic!' she said.

The taxi took off at speed and, this time, I clung to the remnants of an old seat belt for safety. We would soon be risking our lives in the toxic paradise that was Krasnogorsk, but first we were to meet Lyubov: the index case.

In her tiny apartment, Lyubov pulled out her brown envelope of medical notes and showed them to me. She looked exactly like the picture I had seen, except in person she seemed happy. She had sparkled as she met us at the front door of the functional red-brick apartment building in which she lived. It overlooked a much older, more run-down block and a playground that was almost bare, but for one minimalist climbing frame and a swing set. You could tell Lyubov was proud to live in the building that was relatively new and which had flowerbeds along its length.

Lyubov had had the sleeping sickness eight times, but recovered. Her first attack, she told Dinara, happened in April 2010. Her last was in 2014. They stopped after she moved to live in Esil. Now, she felt perfectly well again.

When it began, Lyubov was working in the market in

Krasnogorsk. On one apparently normal morning, another market worker found her asleep at her stall and couldn't wake her. She was rushed to hospital, where she remained in a deep sleep for four days. As the first victim, no one even considered that she might have been poisoned. The doctor's best guess was that she had suffered a stroke. Her brain scan was normal and the symptoms did not fit well with a stroke, but there didn't seem to be a better explanation. Oddly, a few weeks later, Nadezhda, a nurse who'd cared for Lyubov, started to complain of feeling unusually sleepy. As her symptoms developed, she became the second case.

At the time, no one connected the two women's illnesses. That only happened later, when a group of people fell inexplicably ill after the spring festival. They, too, became drowsy and had slurred speech and were unsteady on their feet. There was some speculation that they had been drinking bad vodka, but then somebody remembered Lyubov and Nadezhda. As far as Lyubov knew, that was the first time the idea of an environmental poison was raised. It was also the start of the regular outbreaks.

Dinara and I leafed through Lyubov's medical papers. They described her symptoms at the first presentation: *Answers questions reluctantly; low voice; follows instructions reluctantly; oblivious to surroundings; dizzy.* She was clearly very different then to the effusive woman in front of me, who talked so much it was hard for Dinara to keep up.

Over the course of her eight bouts of illness, Lyubov had every test imaginable, most several times, including six lumbar punctures to look at her spinal fluid. It seemed an extraordinary level of fruitless double-checking, but also a typical Western medical approach to a medical mystery. Tests make patients and doctors feel better. Lyubov had endless blood tests. Hair and nail samples were taken. She also had six CAT scans of her brain. If she hadn't been exposed to excess radiation before, the hospital was making up for it. One doctor said she had cerebral atrophy – in other words, a slightly shrunken brain – but that would not have explained Lyubov's symptoms. Still, the cerebral atrophy worried her and, in her desperation, she travelled to Moscow to see a neurologist. He found nothing wrong and dismissed the brain-scan result as minor and incidental.

'How did you feel in between the bouts of sleeping sickness?' I asked.

'I couldn't stop crying,' she told me, still smiling.

Listening to Lyubov recount her story and reading through her medical records, I found myself constantly comparing her in my mind with Tamara and the doctor I had already met. The one thing that struck me most was that, when she talked about being ill, Lyubov presented it as something in her past. The whole epidemic had felt ever-present, an ongoing source of pain, to the other two. Lyubov seemed happy, while they seemed troubled. There were many common threads to the stories, of course. Lyubov also thought she had been poisoned, although not necessarily deliberately. She thought the source was a pie she had bought from a woman at the market who lived in Kalachi and whose home was very near the uranium mine.

'That might account for the first bout, but what about the other times?' I asked.

She didn't know. The question left her looking perplexed.

We sat chatting in her tiny sitting room, which was also her bedroom. The bed was set back into a wall crevice, a half-pulled curtain partly obscuring it from the rest of the room. Once again, Dinara and Lyubov were doing a lot of talking, leaving me free to look around the room. I caught a glimpse of the corner of a picture hanging on the wall at the foot of the bed, and

I leaned over slightly to see it better. Lyubov followed my gaze and invited me to look at it more closely. I had to stand by the bed and lean in to see it in full. It was an icon and looked very old. It was a family heirloom, her only one, she told me, full of pride. It was her only possession of any value. I was interested to note that she hadn't put this most precious picture in a place where visitors could easily see it. Instead, it was largely hidden from view, but was in a place where she could be assured of seeing it every morning and evening. It was for her own pleasure, not for show.

'I made you food,' she told me proudly, when I had finished admiring the picture, and we moved from the small sitting room-bedroom into an even tinier kitchen.

I'm a doctor, not a journalist; when I listen to stories about illness, I always fall back on a methodical style of medical history taking, even when I know it is not the best way of really hearing what people have to say. In the sitting room-cum-bedroom, I had asked for facts. Once settled in the kitchen, as Lyubov spread the table with plates of food, the conversation became more casual.

'I remember arriving in Krasnogorsk,' she told me, as she served me a milky white soup, which I later learned was made of mayonnaise diluted with sparkling water. 'We were so worried going there, and then we saw it and we were so relieved.'

Lyubov used to live in the Ural Mountains of Russia. Her husband was a miner. In the Soviet era, when jobs were given rather than applied for, he was assigned to work in the uranium mine of Krasnogorsk, and so it was that, in 1975, Lyubov found herself on a bus full of strangers being transported over the Kazakh border to a town she had never heard of before. It was not a journey she wanted to take.

'We didn't know what to expect. I was scared. I remember the bus stopped in a small village and we thought we had arrived. It wasn't a very nice village, so we all panicked, but then the bus started moving again and we were so relieved.'

When the bus finally stopped for good and discharged the people into their new home, they couldn't believe their luck. Brand new blocks of flats, attractive and with every convenience; manicured gardens, trees artfully placed; a river running by. Paradise.

Lyubov was twenty-five years old when she was relocated to Krasnogorsk, largely against her will. Krasnogorsk was a purpose-built colony, set up with the sole function of serving the mine, in an era when uranium was needed and valued. It was also a secret town, she told me. Workers were transported there from Russia. They were given every comfort, presumably to promote productivity and also to keep them in the town and thus protect the anonymity of the place. The 6,500 residents were served by a school, a well-stocked hospital, a cultural centre, a nursery for the younger children, a music school, a fire station and a cinema.

'They had food in the shop that I had never even seen before,' Lyubov recalled. 'Tangerines. They had everything a person could want. Apples. Sweets. Cookies.'

In the 1970s, people in Kazakhstan, and throughout the Soviet Union, lived hard lives. Many food items and other essentials were scarce. Everybody had free healthcare, but it varied in quality. Everybody had a job, but not a well-paid one, nor one they had necessarily chosen. Everybody had a home, but hardly an ideal one. Lyubov's account of Krasnogorsk was helping me to understand Tamara's story. I'd assumed she had been exaggerating the town's splendour. I didn't think any place so isolated could be that wonderful. The pictures of her younger self and the way she talked about her past life made me think she was idealizing that time. But I was wrong. The past had been everything she'd said it was.

'We were under special protection from Moscow,' Lyubov told me.

The residents were given privileges. If a person fell ill beyond the capabilities of the well-staffed local hospital, they were flown to Moscow for treatment.

'When I moved there, everybody in the town was young,' Lyubov told me.

The workers were largely all in their twenties when the town was founded. They started families at the same time and watched their children grow up side by side in relative opulence. The isolation ceased to matter, because they had everything they wanted.

'You could catch fish in the river,' Lyubov said. 'And there was a beach where we could picnic.'

When the Soviet Union collapsed, all privileges were lost. The mine stayed open for a while, but eventually, in the mid-nineties, the Kazakh government closed it, which was the death blow for Krasnogorsk. Many people left; only those who couldn't bear to leave stayed. Luxuries were the first things to go, but soon even ordinary amenities disappeared, with heating and running water eventually withdrawn from many homes. By the time the sickness hit, most of the remaining residents had to collect water from a pump, bringing it to their homes in buckets.

'Why didn't they all leave?' I asked Dinara. 'They had no running water.'

'You've got to understand, the country had just gained independence. It didn't know how to govern itself yet. People were used to automatic jobs and healthcare. Everybody was struggling at that time, not just these people.'

'They had nowhere better to go?'

'Probably not.'

'But, after the sickness started, they were offered new homes. Couldn't they have gone then?'

'I suppose they were used to the hardship, by then.'

If there was a time to fall inexplicably ill, the ten years immediately after the dissolution of the USSR would surely have been it. But the sickness didn't begin until much later.

While talking about all that Krasnogorsk had subsequently lost, Lyubov didn't seem upset. She had weathered the downturn in fortune without complaint and it hadn't dampened her love affair with the place. I wondered if it was because she thought the hard times were temporary. Maybe the people who stayed were waiting, expecting the mine to reopen, the water to be turned on, the shops to be restocked, for everything to go back to how it had been. The structure of the town still stood; nothing irrevocable had happened.

I thought of Lyubov's report that, in between sleeping bouts, she couldn't stop crying.

'Why do you think you cried so much when the sleeping sickness started?' I asked, wondering if something had happened in 2010 that had tipped the balance for these hardy, uncomplaining people.

'I don't know,' she said. 'When I woke in the hospital the very first time, I was crying. The nurse asked me why, but I didn't know why. After that, I cried all the time.'

'Did you feel down, depressed, maybe?'

'I think it must have been gas from the mine getting into my eyes.'

One of my patients with dissociative (psychosomatic) seizures once said to me, 'When I wake from my fits, my eyes are crying.' It's one of the most curious disconnections between tears and sadness I have ever heard. It seems to describe a complete dissociation between mind and body, as if all emotional feeling has been lost. Lyubov reminded me of that.

'You know, most of the time when a person cries, it's because they feel sad,' I offered.

She nodded thoughtfully. 'I remember that, when I moved into this flat, I cried a lot then, too.'

The government had tried to relocate Lyubov many times between 2010 and 2014, but she didn't want to leave Krasnogorsk. The property she was offered in Esil was a fraction of the size of her family home. More than that, she had raised her family in Krasnogorsk and had a deep emotional attachment to the place. But life there was so gruelling that it was becoming untenable, and she kept falling ill. In the middle of all this, her husband discovered he had lung cancer. It was only after he died that she finally agreed to be relocated. The day she arrived in her new home, she sat on the side of her bed and broke down crying.

'I found Esil too grey when I got here,' she said. 'But now I'm planting a garden. Trees will grow here as easily as they did in Krasnogorsk. I am trying to tell the other people that there is no reason we can't make Esil as green as Krasnogorsk, if that's what we want.' She thought for a while, and then added, 'It's true – I cried and cried and cried when I got here. Then I thought to myself, at last it's over, and I stopped crying and I haven't cried since.'

And, with that, all the symptoms of sleeping sickness disappeared.

To best understand the way an illness develops and then progresses, you first need to examine the narratives that have been built around it. Western medicine is not naturally set up to do this effectively, and has a habit of oversimplifying things.

A medical doctor's first impulse is to take a symptom literally. Encountering stomach pain, we are trained to think of bowel disease first. But that is not necessarily what it means to the Miskito. Illness as a metaphor, as a language, as a means to signal distress or conflict, is easily misinterpreted by a system of highly specialized medical doctors working with a list of disease possibilities to go with every symptom.

Even when a psychosomatic cause is suspected, many doctors only have one formulation to explain the phenomenon – blame it on stress. But thinking of psychosomatic illness in that way presents problems, especially when stress is interpreted to assume a single triggering event or a well-defined trauma. Everybody has a source of stress. If you look for it, there will always be some life event or conflict that can be seized upon to explain a person's symptoms. It's an easy formulation. Many doctors, including myself, fall into this trap. I did so with Tamara. No matter how many times she told me that the hardship of life in Krasnogorsk had not made her sick, I would not be derailed from my theory.

The worst outcome for many patients with a psychosomatic or functional disorder can arise when the doctor insists that a particular life event is the cause of the patient's symptoms, and the patient insists it isn't. The two are immediately at odds, with no chance of moving forward. Freud's association between the development of conversion disorders (as functional neurological disorders used to be known) and a history of sexual abuse has been particularly problematic for patients, since it still lingers in the heads of many doctors, and patients can find themselves seemingly accused of denying abuse they know never happened. A proportion of people with functional neurological disorders like dissociative (psychosomatic) seizures have suffered abuse, but a larger proportion have not. Almost as bad is when there isn't an immediately obvious stress-induced trigger, and doctor and patient respond by setting off on a futile hunt for something that either doesn't exist or is too motley to be easily distilled into a simple case of cause and effect. For many people, the development of symptoms is not about a specific traumatic event, but rather is related to embodied expectations, beliefs and stories.

When I read about the sleeping people of Krasnogorsk, I immediately associated the hardship of living in an isolated, dying town with the onset of symptoms. But the real story was so much richer and more complex. For a doctor to help a patient, the details are very important. It was so easy to suggest the people of Krasnogorsk were suffering physically through deprivation, but that did not capture their experience at all.

If one believes, as I do, that the Krasnogorsk sleeping sickness is a psychosomatic phenomenon, then there are a large range of both personal and societal influences that have come together to produce the disorder and propel the outbreak forward. Hardship was not irrelevant, but, as all the people pointed out, they had weathered that without difficulty for more than a decade. It was not the key driver of the illness outbreak. Much more important was the highly unusual town in which these people lived, and their deep – and perhaps surprising – love for that town. Also important was Krasnogorsk's geopolitical position, first in the Soviet Union and later in independent Kazakhstan, as well as the political atmosphere of Kazakhstan and, finally, the media response to the reports of the sickness. Theirs was not a tale of people made unhappy by deprivation, but one about a group torn apart by the difficult realization that they would eventually need to abandon the homes they loved. The sleeping sickness helped them make a very difficult step.

There was no template in the Kazakh culture for sleeping

sickness. It was a finite medical disorder, confined to a small close-knit community. In so-called 'culture-bound syndromes' like grisi siknis, the ideas from which symptoms are drawn exist endemically in the population; in Krasnogorsk, it was created afresh. The scene was set by Lyubov. The first person to develop symptoms in a mass outbreak may have the same symptoms but a different final diagnosis from those who follow. I strongly suspect, based on her description, that all of Lyubov's symptoms were psychosomatic from the outset, but it is possible that it could have been carbon monoxide poisoning, as others have suggested. The group of people who became drowsy and unsteady at the spring festival could have been poisoned. A bad batch of vodka would certainly explain their symptoms. The problem came when Lyubov's illness was connected to the spring-festival incident, at which point a rich narrative began to develop around the illness. With anxiety planted in a community under pressure, others who noticed some bodily change unconsciously looked to Lyubov's experience to know what symptoms to expect. They embodied those expectations.

Whatever my reservations, one of the reasons that 'functional neurological disorders' is a sensible label for biopsychosocial disorders is because it is a constant reminder that the sorts of processes needed to create this sleeping sickness – or resignation syndrome or grisi siknis – are all dependent on physiological processes in the brain over which we have no control. Dissociation is one such biological process and predictive coding is another.

Predictive coding creates real physical symptoms out of expectations that are programmed into our brain networks. It is a physiological and psychological process that is important for our normal daily functioning – but, when it goes wrong, it can also create disability, without need for a brain disease. Our brains don't just accept new information as if inputting it into a computer. New experiences are interpreted based on how the brain has been primed by past learning and experience. The world is just too full of information and sensory experience for us to be able to take it all in and interpret it afresh every time. Instead, through unconscious mechanisms, extraneous information is filtered out (for example, the feeling of clothes against your skin) and what remains is assessed according to experience (I can make it across the road before that car gets to me).

Visual processing provides a good example of how the top-down processing system involved in predictive coding works. As sensory information enters, it is being compared to our expectations drawn from existing knowledge and is being manipulated to fit with those expectations. As you look at a scene, the visual-processing centre adds and subtracts elements. It helps you focus on the object of your attention and filters out things that are unimportant in the moment. It compares what you are looking at with objects familiar to you. It judges speed and colour and depth. So, as visual sensory information enters from below, our brains are working at a higher level to make sense of it. Our brains manipulate images so that what we see is not faithful to the scene, like a photograph, but is instead an interpretation of the scene. That is the reason we can read lots of different styles of handwriting. The brain puts the various swirls and shapes in the context of a sentence, compares it to the known alphabet and makes a best guess. Top-down processing is why you can read nonsense sentences like this:

YoUR M1ND 15 R34D1NG 7H15 4U70M471C4LLY W17HoU7 3V3N 7H1NK1NG 4BoU7 17.

Predictive coding has been used as a model to explain resignation syndrome. It suggests that Nola and Helan and the other children have prior expectations coded in their brains, telling them how their bodies will behave in response to a particular situation. At some conscious or unconscious level, they knew that children faced with deportation can become listless and fall into a coma. It was inevitable that the asylum process would trigger an emotional reaction accompanied by a fight-or-flight response, with all the physical sensations that entails. Their brains were primed to shut down as soon as the first physical effects of their circumstances were felt. As Karl Sallin, a doctor researching the disorder, suggested, 'overwhelming negative expectations led to a down regulation of behavioural systems.'

Neuroscience often refers to the brain's coded expectations as templates or priors. Priors make us more efficient and allow us to negotiate the world more easily. They serve an important function. However, they are not necessarily always accurate, and therein lies the problem. They are a well-educated guess. If the guess is wrong and the input signal doesn't match the priors, there has been a prediction error. That offers a dilemma with two possible solutions. The brain may alter the priors to incorporate the new incoming signal, and therefore learn from the new experience. Alternatively, the brain may process the incoming signal to fit with the priors.

Think of it like this: you meet your neighbour in the street and they say something to you. Your neighbour is usually rude, but, on this occasion, they seem pleasant, so you adjust your view to incorporate the idea that they can be nice sometimes. The prior expectation has thus been altered to accommodate a new experience. But there is another way the scenario could play out. When the neighbour is uncharacteristically pleasant, the brain might not accept this as prediction error, but could instead alter the interpretation of the input signal to bring it into line with expectations. You find insult in what the neighbour said, because you know that's all they're capable of. The brain could therefore process the same experience in two entirely different ways, and it is believed to do this with every type of sensory input. The same physical touch could feel different depending on the context and the associations – how you are feeling that day, who is touching you and why. The brain commonly alters a person's experience of sensory stimuli unbeknown to them. Obviously, a person's general well-being at any given moment will have some impact on which route the brain will take. When a person is feeling vulnerable, they are more likely to reach for the negative end result. In the example above, if there were stories going around about that neighbour on that day, it would influence the interpretation of the input information, too.

Inaccurate expectations may be an important feature in the development of functional disability. Priors affect how we interpret and react to bodily changes. Take as an example a person who once developed severe laryngitis in association with a viral infection and lost their voice as a result. When that person notices the next sore throat, their brain will process the experience by comparing it to priors. A prior template error could tell them that every time they get a cold, they will lose their voice. The top-down priors overwhelm the sensory input and induce them to temporarily lose the ability to speak. This is an unconscious brain process.

If physiological processes like prediction errors are indeed responsible for creating the disability of psychosomatic and functional disorders, that should not be understood to mean they are just biological conditions and that psychosocial vulnerability, personal conflict and social influences are irrelevant. They are often the trigger. When faced with conflict or unhappiness, we may notice physical changes, act them out based on templates and use them as a means to ask for help or to solve a problem. We are more resilient to illness when things are going well for us.

As discussed, our expectations of health and ill health take the form of illness templates, coded in our brains by our sociocultural environment. The development of sleeping sickness could therefore be explained by the embodiment of social narratives and the unconscious enactment of predictive errors. Lyubov's illness and the outbreak at the spring festival were distinct events in themselves, but, once the rumour that the town was toxic began to spread, people searched for symptoms in themselves. When you look for symptoms, you find them. Imagine you ate some food that you later learned was prepared in a restaurant that was closed that day for being unhygienic. It would be natural to look at your own body for signs of food poisoning, and to feel nausea just from the idea. The people of Krasnagorsk, being concerned about poisoning, searched their bodies and played out the story predicted for them. But, because stories change from person to person, everybody who was dragged into the outbreak brought a new element that expanded the symptom pool. Disease symptoms are very similar between people, but biopsychosocial illness evolves. Predictive coding is also likely to be very important to grisi siknis, but, as a culturally embedded disorder, it has a stable template dating back decades. That made the symptom pool more consistent between individuals – although, not entirely consistent, as the different accounts from different generations showed.

Psychosomatic disorders can be constructed from internalized stories. The narratives are built like scaffolding, unstable at first, but, stage by stage, new elements are added, until the construction is sturdy enough to sustain the symptoms. The scaffolding supports mistaken beliefs about medical conditions and makes them feel unshakeable to those affected. This is a complex, sophisticated process, which usually serves an important purpose. Attributing a person's symptoms purely to stress, and failing to appreciate the nuance and complexity, can make the patient feel as if the doctor is taking a sledgehammer to the scaffolding. The patient is compelled to reinforce the scaffolding or to crumple.

After our meeting with Lyubov, Dinara and I stayed overnight in a hotel in Esil. We were booked to take the train back to Nur Sultan the following afternoon, but not until we had visited Krasnogorsk. I almost missed that opportunity when some twilight-zone feature of Esil reset my phone to Russian time overnight, while Dinara's stayed faithful to Kazakhstan – another reminder, perhaps, of the history of the region. As a result, I woke very late and didn't get to have the pancakes for breakfast that Dinara assured me were the best she had ever eaten. Our pre-booked taxi driver was already waiting, and looked impatient to leave.

The car that took us to Krasnogorsk was sturdier than our previous transport. It wasn't a long drive, but the road was exceptionally rough and potholed. In places, it was so bad that the driver had to pull off the road and drive through a field to avoid the impassable craters in the dusty red earth. We were driving through the steppe – a vast, flat grassland, coloured green and straw yellow, that stretches from northern Kazakhstan into Russia.

'This used to be a really good road,' the driver told me, via Dinara.

The best road in Kazakhstan, I thought to myself.

'He says it was the best road in Kazakhstan,' Dinara added, and started to laugh. She had read my mind. 'He says that people who could afford it used to drive for hours, just to go to Krasnogorsk to shop. It was so well stocked, people could get things there that weren't available anywhere else in the country. Then the mine closed, the shops closed, and the road was allowed to disintegrate.'

There was little to see along the way, except for hillocky grass. Every few minutes, I would notice a gopher pop its head up from a hole and watch us go by. Neither Dinara nor I had ever seen a gopher in real life before, but, here, we were treated to hundreds of them, appearing and disappearing every few seconds.

'Maybe they aren't gophers. Maybe they're actually irradiated mice,' I suggested, still musing on the uranium mine. Wisely, perhaps, Dinara chose not to translate that for the driver.

Then, there it was, myth made real. Krasnogorsk rose up from the bleached, faded steppe, and it was not a disappointing sight. From a distance, it looked like modern apartment blocks embedded in green trees. The trees were a much deeper green than the grassland, which made for a stark contrast, and it reminded me of a post-apocalyptic scene in a movie, in which one can appreciate the grandeur of how the buildings once were, even though they have become overrun with jungle.

Kalachi lay right beside Krasnogorsk, with a single road dividing the two. I hadn't appreciated before that they were essentially the same town, only separated by age and architecture. Kalachi was a farming town with low-level buildings, compared to the high-rises of Krasnogorsk, and it had existed long before Krasnogorsk was founded. It had got lucky, for a few decades at least, by having paradise placed on its doorstep.

As we got closer, the scene changed. The apartment blocks on the Krasnogorsk side turned out to be mere skeletons of buildings. The windows were glassless and frameless, and looked like big empty eye sockets. The doors were gone. The roofs were stripped. One or two of the buildings were crumbling into piles of powdered bricks. The gardens that Lyubov had told me about ran wild. We drove in between the buildings. The further into the town we got, the worse the road became.

'People lived in these five years ago. How can they be this decrepit already?' I asked.

The driver told us that scavengers had picked the vacated apartments clean. Anything of any possible value was ruthlessly hacked free. Tamara and Lyubov's apartments were rubble now.

'It's still beautiful,' Dinara said.

I agreed. Maybe it was the blue sky against the green foliage, but you could still see the beauty behind the destruction. We stopped and took photos, and I tentatively entered one of the buildings. The floors had been pulled up and piles of discarded rubbish lay everywhere. The cultural centre that Tamara had worked in was equally torn apart. We called into the once-thriving hospital, which was not only still intact and well preserved, relative to the other buildings, it was also open. Of the hundreds of staff employed at the hospital in its heyday, only one nurse remained. This one woman served the whole community, approximately 300 people between the two towns. Dinara had phoned ahead to request an interview. Like the doctor in Esil, she had been a little reluctant, and Dinara had to ring her a few times to secure the meeting. Still, the nurse greeted us warmly, and proudly showed us around her domain. The hospital was large, but only a few areas were still in use. A bath filled with water in the sluice room provided a reminder that this was a town that could not rely on its water supply. There were no inpatients, just cavernous rooms with empty beds waiting for transient visitors. The seriously ill would be taken to Esil.

The nurse invited us into her office, where there were 300 browning yellow packets of letters that contained the medical histories of every patient, all of whom she knew by name. The nurse was the only medical support and she was nearing retirement age. When she leaves, I thought, they will never find a replacement, and that, surely, would sound the ultimate death knell for Krasnogorsk and Kalachi.

She talked about her attachment to the town. She had once been a surgical nurse on the trauma team, but, as more and more staff left the hospital, she had become a master of everything. She had been the first port of call when people started to fall asleep.

'They were like people waking from anaesthetic, although everybody woke differently. It changed every time.' She echoed the stories of the others and told us that, when tests were normal, she had given the people intravenous fluids to flush out their system.

When I asked her if she would consider a psychological cause, she laughed and said, 'But how would a psychological problem affect a child?'

I asked her if she would ever leave Krasnogorsk, and she said no. It was her home. Seeing the place for myself, and having talked to Lyubov, I understood her position so much better than I would have before.

After we said goodbye to the nurse, Dinara and I wandered the streets for a while. A small number of families still lived there, having refused to leave. We had arranged to visit one home that was still occupied by a husband and wife in their late fifties, and the wife's elderly mother. They had a detached bungalow, one of the nicest houses in the town. It took prime position, with a view of the river, and was surrounded by flower beds and a vegetable patch. The wife told us that she didn't want to swap her three-bedroom house with private garden for a tiny upper-floor apartment in Esil. She was holding out for better. Her husband, meanwhile, was perfectly happy to stay. He saw no need to relocate; he wasn't sick, the quiet of the town was to his liking and he could still fish.

The woman told me she'd had a phone call from the government just before Dinara and I arrived. 'They're watching you,' she said. I doubted this was true. I had media accreditation to make the trip, but I hadn't been asked to specify an exact date of travel. I hadn't pre-booked the train or hotel, so it was hard to believe anybody knew or really cared I was there. The phone call was about the woman's compensation case and she was convinced they had only phoned that day because of our meeting. I was inclined to dismiss her comment as paranoia, but I have never lived in a country that blocks social media and access to news websites, so I couldn't be sure. I had certainly stood out as a foreigner in Esil.

We spent the rest of the morning absorbing the atmosphere. The buildings had been reduced, but the plant life had been left untouched. It had flourished, out of control. In some places, fallen trees and overgrowth blocked the road. We meandered between fallen buildings, stumbling over rubble and shrubs. Every now and again, we came across men who made their living from entering these buildings to find anything of value that had been missed. One told us he had the sleeping sickness. He was dressed in dirty workman's clothes and he rested his chin on the handle of his spade as he asked me why I was visiting. When, in turn, I asked him about his experience of the sleeping sickness, he wouldn't answer any questions about it. He shuffled away at speed when I took my camera out. If anybody else was interested in my visit and was watching me, they were well hidden.

In the afternoon, with nothing left to see, we renegotiated the potholed road back to Esil. The gophers that had so keenly watched us arrive were all gone. My final few hours with Dinara were spent on the train to Nur Sultan. Because the distances that people travel in Kazakhstan are so huge, nearly all the trains have sleeper compartments; we were sharing a four-berth compartment with two Kazakh women. We lay in the top bunks, talking across the divide, and they lay in the bottom bunks, doing the same. You couldn't see out of the window from the bunks, but I knew what was there – flat grassland, pylons, the occasional grey town.

'What do you think of it all?' Dinara asked me.

'I think I had it all wrong. When I read about the sleeping sickness in the newspaper, I just assumed the people were falling asleep because the town they lived in was so bleak and the illness made their lives exciting.'

'Because they had nothing else to do?'

'Sort of. But now I see it differently. They are tough people. Nobody we met complained about how difficult life had been in Krasnogorsk, not once. Why did it take me so long to hear them say that? They only complained about having to leave. They didn't get sick because they were unhappy with their lives; the problem was their love for the town and how exceptionally it had served them for so long.'

'They raised their families there.'

'Exactly; they had so many good reasons to have a deep emotional attachment to Krasnogorsk. After independence, when it went downhill, they might have lost their luxuries, but they never lost their romantic connection to the place.'

'You can see why they clung on for so long.'

'I think they knew, on one level, that they would eventually have to leave, but they didn't want to. The sleeping sickness helped them make the difficult decision that had to be made. It excused their leaving.'

We lay there thinking about that for a while. What I didn't say was that I thought the doctors' over-exuberant testing had probably made things worse. The journalists played a role, too, in exacerbating the problem, along with the government response and the political climate. Lyubov's illness was an inciting event, a template, but fundamental too were the unique lives of the people of Krasnogorsk. The repeated medical tests reinforced the paranoia about a poison. In Krasnogorsk, psychological services were neither available nor offered, but brain scans and lumbar punctures were done liberally. Scientists came to the town to investigate the environment. The reaction to physical symptoms was rigorous, which reinforced the disease narrative and invited new people to look for symptoms. Telling and retelling the story to journalists reinforced the fear and intensified the search for evidence of illness. Journalists duly strengthened both the concern about the poison and the stress-related paradigm to explain the illness. People were pictured looking despairing, standing among the rubble of ruined buildings, beside captions that read, *If the residents of Krasnogorsk and Kalachi were exhausted, they had plenty of reason to be.* That over-simplified portrayal of their lives and the town they loved cornered the residents into a position of having to fight even harder for a disease-related explanation for the sleeping sickness.

Krasnogorsk had the opposite effect on Dinara and me.

Dinara finally spoke: 'I feel astonishingly relaxed.'

'I do, too. Maybe a big dose of radon gives you sleeping sickness, but a small dose, taken with the morning air, just chills you out.'

All the way from Nur Sultan to Esil, we had been speculating about the wisdom of going to a town with a suspected radiation leak.

'Lyubov was so positive,' Dinara said. 'She had the smallest apartment, but she was the most content.'

'I agree.'

'You know what the name Lyubov means in Russian, right?' Dinara asked.

'No. What does it mean?'

'Love.'

4

Mind Over Matter

***Physiology:** the way in which a living organism or bodily part functions.*

Immediately after we left the home of Nola and Helan, in Sweden, Dr Olssen took me to a first-floor apartment in the same block, to meet Flora and Kezia. Her husband Sam and the family dog came with us again. On the way up the stairs, I asked Dr Olssen if the two families were friends. They hadn't known each other before the two sets of girls fell ill, she told me. The second family had only recently been moved to the apartment block by the authorities. The girls sometimes met in the playground when their parents brought them outside in wheelchairs to get sun and fresh air.

'These girls are Romany,' Dr Olssen told me. 'From Albania. They have nothing in common with the others. They speak a different language.'

That is not strictly true, I thought. They might have come from different places, but they have a great deal in common now.

I recognized the girls as soon as I saw them. A fifteen- and a sixteen-year-old, they had been pictured in one of the newspaper stories that had led me to Sweden. They had been in resignation syndrome for five years. Later, when I reread the news article in which I had seen their photograph, I found their nationality was listed as Kosovan. Their ages differed slightly from what I had been told, too, and I supposed the discrepancies could be something to do with both the nomadic nature of the Romany people and the tumultuous history and disputed geography of the region. Or perhaps this was just another example of the blurring of identity that seeps into the lives of anyone forced to flee one country and seek asylum in another. These girls and their family had left everything behind.

Their beds were arranged in an L shape, pushed up against two walls. The similarity of the scene to the newspaper photograph, taken over a year before, was eerie. It made it seem as though they hadn't moved, or been moved, since that time. They lay under pretty floral bed covers. Like the younger girls in the flat below, they each had dark hair that made a halo on their pillows.

At Dr Olssen's urging, I examined them. When I lifted Flora's eyelids with a tentative thumb and finger, they opened easily to reveal glassy eyes staring at the ceiling. I'd had the sense that Nola was aware of her surroundings – I thought that maybe she didn't want to see me and was averting her gaze – but these girls gave no sign of knowing I was there. Flora's skin was pockmarked with adolescent acne. She had been a child when she fell asleep, but, as she lay there, her body was maturing. Both girls' feet were stone cold, despite the warm day, and the skin of their fingers and toes had a slightly purple tinge. They were well cared for, their joints moved easily and their skin showed no signs of the ulceration caused by immobility, but the discolouration was a sign of poor circulation. People need to move to keep the blood flowing. These two looked pale and unhealthy – not restful, but sick. I feared I was looking at what the future held for Nola and Helan, if they didn't get the help they needed.

Flora and Kezia's family had been refused asylum a few years before. Their parents were terrified at the prospect of taking the family back to their country of origin. Romany people are a persecuted minority – they were sold into slavery in the nineteenth century, and, during the Second World War, the Nazis sent them to the concentration camps and gas chambers. Josef Mengele, the Angel of Death, was said to favour Romany children for his inhumane experiments. In the 1980s, in an attempt to limit their number, Czech authorities subjected them to forced sterilization. Later, in the Yugoslav conflict of the 1990s, the Roma were caught between two sides. They were ultimately seen to affiliate themselves with the Serbs, but, according to media reports, were subjected to torture and abuse from both warring parties. When the war ended, the Kosovo Albanians razed Romany districts to drive them out. 'We have no country to go back to,' the girls' mother told me.

As I moved the girls' limbs to assess their tone and the health of their joints, they remained oblivious to my presence. Their brother watched me from the doorway. He had a Swedish flag painted on each cheek and a Swedish scarf hung around his neck. He was waiting for a football match to start, decked out as if he was in the stands. Dr Olssen asked the boy if he felt okay. He replied in Swedish and she translated, telling me that he was complaining of headaches and dizziness. She talked to him again and he said he was sleeping badly and having nightmares. I wanted to tell her to stop asking him about symptoms; he lived in a confined space with two chronically ill sisters and, since illness is passed on through proximity, expectation and embodiment, I was worried for him.

After I had pulled the girls' bedcovers back into place, I was led into the sitting room, where a large television was showing the pre-match commentary for the Sweden versus England World Cup quarter-final. A table was laid with food. There were bowls filled with peaches, cherries, apples and mandarins, and I was offered potato patties, which I accepted, following my hosts' example by sprinkling them with salt and paprika. We didn't talk about resignation syndrome – we watched the football. The boy sprawled on the floor, staring at the television, only turning towards us when his mother finally started cutting the cake – a chocolatey, multilayered affair, filled with buttercream. The atmosphere in the room got competitive when, thirty minutes into the game, an English player scored. I was obviously the only one shouting for England. I had lived in England for more than ten years before I began supporting my adopted home in competitive sports. This family, their sitting room bedecked with Swedish bunting, had got there much quicker. Clearly, why you left and what you left behind makes a difference.

At half time, Dr Olssen, Sam and I prepared to say goodbye. As I walked through the hallway towards the front door, I caught a glimpse of Flora and Kezia lying just as we'd left them an hour before. I suddenly felt very guilty. While eating cake and watching football, I hadn't thought of them at all. Life went on for the rest of us, while they just lay there. They had gone to bed as children, but when – *if* – they woke up again, they would do so as adults, with very different bodies. They would have a teenage brother, who had barely started school when they last saw him. I just hoped that they would wake up in Sweden.

Most people agree that the resignation-syndrome children are suffering through loss of hope. It was difficult to know what might be ahead for Flora and Kezia after so many years of apathy. Is it possible they will never wake? I find it hard to believe that is their future. I have seen many people's lives destroyed by psychosomatic illness, but I have never seen loss of life. Then again, there is some precedent for death brought on by hopelessness. It's exceptionally rare, but rare does not mean it never happens.

Primo Levi wrote about this in *If This Is a Man*, his book about his time in Auschwitz. In the Nazi concentration camps, he said, people could be divided into the saved and the drowned. He believed that a similar division also occurred in ordinary life, although it was less evident; the hopelessness of the camps brought it out. The drowned were only in the world on a visit, he wrote, and would be turned to ash very quickly. Those people who resigned themselves to death were referred to as *Muselmänner*. They ate what little they were given and they followed orders. But they did these things with expressions that were lifeless, long before they died. 'One hesitates to call them living, one hesitates to call their death death,' Levi said of them.

A *Muselmann* was a person who had lost the will to live. Theirs was an extreme hopelessness, a willed death. Of course, it's not hard to imagine a person giving up to the point of death when they are malnourished and tortured and completely deprived of humanity and of hope, as people were in concentration camps. It's harder to imagine that a person could wish for and get death without the pre-existing physical frailty, and in less extreme circumstances. But something very like that happened to a group of Hmong refugees from Laos, in the USA, in the 1970s and again in the 1980s.

The Hmong people are an ethnic minority originally from China, many of whom fled to South East Asia to escape persecution at the end of the nineteenth century. During the Vietnam War, the USA recruited Hmong people to fight against the Russian-supported Lao troops. When the Americans abandoned the area, many Hmong people fled to the States as refugees. Within a year, dozens of them had died in their sleep for no identifiable reason, with no warning or preceding illness. The men appeared to be in good health, but they went to bed one night and never woke up.

Sudden unexplained death can occur to people of any culture and ethnicity, but – for a period, at least – it had a much higher incidence in Hmong people. At the time, the Centres for Disease Control and Prevention speculated that the men had died from cardiac arrhythmia, but couldn't say why it had happened. One could certainly speculate that there's a genetic reason – perhaps they shared a genetically driven heart condition. If that was the case, though, something similar should have happened to Hmong people in other countries, but it didn't. Some blamed the chemicals used in warfare, but the Hmong were not the only people exposed to that war and those toxins. A decade after it started, none of the medical theories could make sense of it.

With no better explanation, many wondered if the deaths were linked to the stress of trying to assimilate into North American culture. The Hmong immigrants of the time were illiterate and they didn't speak English. They were used to living in the mountains and their kinship structure was polygamous. In the USA, large family groups were forced to live in confined spaces, they didn't know how to use modern appliances and they had little hope of getting work. The life of a forced immigrant was a huge struggle. There was widespread speculation that they just gave up – they decided to die.

The Hmong people had their own thoughts about what caused the deaths, however. They were a spiritual community who believed that death could be caused by evil spirits. They believed that the soul could quite literally be shocked out of the body. All of the men had died during sleep or as they fell asleep.

Some deaths were witnessed, and the victims were seen to moan and cry out just before they died. The Hmong suggested that they had been terrorized to death by a nightmare. Nobody has ever offered a better explanation. It is a true mystery.

Other cultures also believe that death can be willed or can occur for magical reasons. The Aboriginal people have a traditional practice called 'bone pointing', in which an enchanted bone or spear or stick is pointed at another person to cause their death. The victim is said to die within a month, with no other cause found. In Haitian cultures and among the Maori of New Zealand, a similar belief is referred to as 'voodoo death'. These are deaths caused by an expectation.

Most Western people do not subscribe to that brand of spirituality and therefore death by will or curse or nightmare seems strange and unbelievable. But we have plenty of our own stories of people who have died simply because they wanted to. Almost everybody has heard at least one account of a previously well elderly person passing away within hours or days of their beloved life partner. Terminally ill people are often said to 'give up'. They take control. They decide it's time and succumb to death. Stories like these are just anecdotes. It's impossible for most people to will their own death, no matter how much they want it. But is the Hmong belief that a nightmare could cause death so absolutely implausible? It is certainly the case that shock can cause measurable physical changes, changes that have the potential to lead to sudden death. It is entirely possible for a person to die of a broken heart. It doesn't happen often, but it happens. In 2018, Karin, a woman in her late fifties, a previously fit and well US resident living in Pasadena, California, learned that the hard way. She didn't die, but what happened to her made death feel frighteningly close, for a short time at least.

Karin was a person accustomed to good health. She was a vibrant member of society. A business woman and a philanthropist, she was always busy. She worked as an executive director of a non-profit charitable organization providing childcare for low-income families. In her spare time, she was a keen horse rider. She told me her story over the phone, and her youthful voice made her sound much younger than she was. When I asked her to describe herself, she said she was a five-foot-three tomboy.

'I like outside stuff more than I like inside stuff,' she said.

Right up until the day she had her near-death experience, Karin had felt as invincible as we all do when we have never known a serious health problem. She'd had some minor medical complaints, but never any of the life-threatening variety. Her medical history mostly involved bumps and scrapes related to horse-riding accidents, as well as a number of operations for a retinal problem. It was during one such routine operation that her crisis occurred.

On the day in question, Karin had been entirely unconcerned about the procedure, which did not even require admission to hospital. She'd had the same treatment carried out under similar circumstances before, so she went to the appointment alone. She was relaxed and knew what to expect. She was sedated, but not anaesthetized – therefore, as the procedure began, she was not asleep and had a vague awareness of what was happening around her. She could hear the anaesthetist chatting with the surgeon. She tuned them out and let herself daydream.

But this operation would not prove to be as routine as those she'd had before. A few minutes in, Karin was jolted out of her reverie by a sharp change in the tone of the anaesthetist's voice. Nothing like that had ever happened at any of her previous appointments.

'I could hear his voice rise and then he suddenly started yelling at the surgeon. At almost the exact same time, I felt like the top of my head was going to blow off,' she told me.

Karin heard the anaesthetist shout at somebody to give her epinephrine (adrenaline), after which she lost consciousness. For a while after that, she drifted in and out of awareness. When she woke fully, a lot of time had passed and she was surprised to find that she was no longer in the operating theatre. When she was told that she had needed to be resuscitated and rushed to the intensive-care unit, she was horrified. She learned that her blood pressure was dangerously low and the doctors were struggling to treat it. There was something wrong with her heart.

'I'd never had a heart problem. Even when it happened, I didn't have any pain in my chest,' she told me, still sounding astonished.

Halfway through Karin's operation, and with no clear provocation, her blood pressure had crashed. She had been given rescue drugs, but it had taken over an hour and a transfer to the intensive-care unit before she was out of danger. As soon as her blood pressure was stabilized, she was transferred to the radiology department for an emergency cardiac angiogram. Karin was awake for the procedure, during which she heard the cardiologist express surprise that there was no evidence of any atheromatous plaque. He had expected to see signs of a blockage or prior heart disease, but the problem was not lack of blood supply due to furred arteries; she had not had a myocardial infarction (heart attack). However, they had found something amiss. Karin could hear an odd change of tone in the doctors' voices, stirred by whatever they could see on the radiologist's monitor.

'I could hear them talking. They were all leaning in, amazed by something. New people kept coming to have a look. By the end, there were maybe twenty-five people in the room. I heard them say they had never seen a heart like mine before.'

She spent the next six days in the intensive-care unit. A machine alarm kept sounding every time her blood pressure dropped. The first night, she was sure she would die. She didn't. With careful nursing, she recovered.

'I almost died,' she told me.

'But you didn't,' I reminded her.

'I know, and I've learned from it. I'm going to listen to my body, in future,' she said. 'I thought I'd have to sell the horses.'

'Are you putting it behind you, now?'

'I am. Although, I live by a toddler's schedule – having naps and working short days.' When she laughed, she sounded very young. I got the impression that the health scare had helped her to make some positive changes in her life. It was a good excuse to do more of what she loved and less of what had made her sick in the first place.

Karin's angiogram and subsequent echocardiogram (ultrasound) showed that her heart had blown up like a balloon. She had no pre-existing heart problems or family history of heart disease. She'd had several heart tracings done as part of routine preoperative assessments in the past, and they had always been normal. Health checks with her family doctor had always come up clear. She'd had the same operation and the same sedating drugs in the past, without difficulty. There was nothing to suggest she had been harbouring a heart problem for some time. What happened to her in the operating theatre that day showed all the signs of being something abrupt and brand new to her.

Karin was diagnosed with broken-heart syndrome. The doctors told her that stress had caused acute heart failure, a condition referred to in medical circles as takotsubo cardiomyopathy, in which the heart muscles become suddenly weakened. The left ventricle wall bulges and contracts abnormally and the whole chamber changes shape, which means the heart can no longer effectively pump blood around the body and the blood pressure falls dramatically. If it isn't treated promptly, it can cause sudden death. That it happened to Karin in a hospital was probably what saved her.

Takotsubo cardiomyopathy is poorly understood, but it classically occurs at the time of a sudden emotional or physical shock, or during periods of severe chronic stress. Typical triggers are bereavement, serious illness or accident, violent arguments, intense fear or financial loss. Some reports also describe it as being triggered by the anxiety of public speaking, or even by a surprise party. Scientists' best guess is that it is caused by a surge of stress hormones – adrenaline, in particular – which stun the heart, causing it to stop contracting effectively. There are many unanswered questions, particularly why it should be so much more prevalent in women. Since it almost always happens after the menopause, it has been suggested that middle-aged female hearts are made vulnerable by a fall in oestrogen levels.

'I was under enormous stress for two years before it happened,' Karin told me.

When Karin told me about her work, I thought it said a lot about her. Decades before, she had set up the charitable organization that she worked for. The idea arose from a casual chat with friends, after church one Sunday. One moment the group was sympathizing with the plight of low-income families who had difficulty accessing childcare, the next they were setting up a service to solve the problem. She was the sort of woman who was always problem-solving and doing things for other people. But she noticed that, as she got older and stresses accrued, she was beginning to struggle. Financially, she was fine, but her work was becoming increasingly demanding. Added to that, her parents, who lived hundreds of miles from Pasadena, were sick: her father had developed dementia and her mother had Parkinson's disease. Karin wanted to support them, but the distance made it difficult. She was concerned that they were not well enough to continue living independently, so she started looking for a nursing home, near Pasadena, where they could still live together and she could see them regularly.

'Can you imagine how hard it is to find a facility that will take one person who is physically able, but is a smoker with mental problems, and quite a difficult person, *and* a second person, who is mentally sharp, but physically disabled?' I could hear the frustration, but there was also lots of humour and fondness in her voice as she described the impossible feat.

It took two years for Karin to find a suitable facility and then to get her parents to agree to go there. This she did while also caring for her own family and managing a business. A week before they were due to relocate, her parents called off the move, telling her they had changed their minds. All her effort came to nothing.

'Sometimes I wanted to scream, but I wouldn't let myself. All the time, my workplace was getting tougher and tougher. I couldn't stop thinking about my parents. I had a constant feeling of unease in the pit of my stomach. Now, I know that feeling was epinephrine [adrenaline].'

Not long before Karin's collapse, both her parents died – first her father, then, only two months later, her mother. Meanwhile, her job was relentless. Some days, she got dressed for work, but could barely make herself leave the house.

'I'd find myself standing, staring at the front door, not able to open it.'

Until, eventually, during a routine procedure, her heart expanded like a balloon and threatened to pop.

'I had never heard of takotsubo cardiomyopathy until I got it,' she told me. 'I assumed it was called after some famous doctor. I thought there must be a Professor Takotsubo out there somewhere. I could only laugh when I heard it was actually named after an octopus pot!' She laughed again at the memory.

Takotsubo cardiomyopathy was first reported in Japan, in 1990, and was indeed named after a distinctively shaped pot used to trap octopuses. When the heart expands, it takes on the shape of the pot.

'How do you feel now?' I asked her. It had been a year since her heart problem began.

'It's still difficult. I enjoy hiking, but I get breathless walking uphill. But, more than that, I have to be careful with what I tell people. If I say I have stress cardiomyopathy, they are immediately less impressed by what I've been through – as if anything to do with stress is less deserving of sympathy. Insurance companies are also less likely to pay for physical therapy if they think the problem is due to stress. I don't even have a scar to show for it.'

It is sad that, even when someone's heart stops working and actually threatens their life, they cannot let stress seep into the conversation for fear it will detract from the reality of their experience. It is also typical that, without a lasting scar or visible physical disability, others find it quite easy to forget what happened.

'I tell myself things are not the same, but they are still going to be okay.' She sounded pragmatic. Something bad had happened to her, but she was gathering herself, preparing for how to deal with this new normal.

'They will be okay,' I assured her.

She started laughing again. 'I actually blew a gasket!'

If you look up 'organic brain disease' on Wikipedia, it is defined as 'any syndrome or disorder of mental function whose cause is alleged to be known as organic (physiologic) rather than purely of the mind.' I am not advocating that anybody get their facts from Wikipedia, but rather highlighting how easy it is to find evidence of dualism in the twenty-first century. This definition suggests that the mind has nothing to do with physiology, which is obviously ridiculous. But Wikipedia is far from alone in dividing medical disorders into 'organic' and 'psychological', as if the two could be divided so absolutely in that way. When you have a thought or feel an emotion, something organic (physiological) is happening in your brain to create it.

In the medical field, the term 'organic' is generally used to refer to a pathological change in an organ, while 'non-organic' refers to a disorder that has a psychological cause. I don't have strong feelings of disagreement when those definitions are used as guidelines; however, I recognize they have become problematic because many people, both medical and non-medical, extrapolate the 'organic' and 'non-organic' division into meaning that symptoms are either 'real' or 'not real.' Therefore, someone who has a stroke is referred to as having an organic brain disease, and they are considered to be 'really' paralyzed, whereas somebody who has a psychosomatic or functional disorder is considered to have a psychological problem, and so they are not 'really' paralyzed. With that sort of interpretation of what it means to have a psychosomatic (functional) disorder, is it any wonder that people reject the diagnosis?

Whatever you call these disorders – functional, psychosomatic, biopsychosocial, conversion, non-organic – they all arise as a result of physiological mechanisms that go awry to produce genuine physical symptoms and disability. Although they are often explained by a simple cause-and-effect model between stress and physical symptoms, there are actually a multitude of mechanisms through which they can develop. They are a manifestation of the interplay between the body and the higher cognitive and social processes that create the 'mind'. While there may be no disease present in terms of a pathological structural change, there are physiological abnormalities created by such brain and bodily functions as predictive coding, dissociation, stress hormones and the autonomic nervous system.

Karin didn't have a psychosomatic disorder, because she had a definite structural change in her heart, which, by current definitions, makes her problem organic. However, what her illness demonstrates is the very complex interaction between the body, the cognitive processes that make up the mind, and social pressures. In Karin, palpable stress produced a dangerous surge of hormones that produced a life-threatening cardiovascular response. There is no doubt that psychological distress produces physiological changes; it stimulates the autonomic nervous system and activates the amygdala, the brain's early warning system, inducing the release of cortisol and adrenaline. This is a normal physical response to acute stress and is supposed to help us prepare for fight or flight in the face of imminent danger. Because high levels of these hormones can cause problems in the long term, such as raised blood pressure and heart disease, they are subject to modulation by a feedback loop under the influence of the hypothalamic–pituitary axis. The purpose of the feedback system is to try to dampen the magnitude of the stress response, if stress becomes chronic. Clearly, the feedback loop didn't work very well for Karin.

But the interaction between mind and body doesn't have to start with psychological distress; it is a two-way street. Functional neurological disorders and psychosomatic disorders are created by traffic in either direction and are exacerbated by feedback loops between the two. In Karin, the psychological provocation came first, the physical consequences came later. What happened to her fit well with the traditional model through which psychological factors are thought to effect physical health. But there are so many more ways for this interplay to work. The psychological contributors to illness may be negligible at the start. That is how it was for Tara, a patient of mine. Her story began five years ago, with the worst pain she had ever experienced, which came to punctuate an otherwise happy life.

'Giving birth is less painful,' she told me, when I met her in the outpatient department.

She was referred to me after she collapsed, so I was surprised to see her arrive in a wheelchair. There was obviously a great deal more to her story than the single blackout described in her referral. In fact, even she was not concerned about the collapse. She attributed it to a faint and thought a referral to a seizure clinic unnecessary. When she described her collapse, which occurred in the context of a sudden piercing pain in her back, I agreed with her self-diagnosis. I asked why she had come to clinic, if she was so unconcerned by what had happened.

'Because the doctor said they would take my wheelchair away if I didn't come to see you.'

Clearly, there was a lot I hadn't been told. I asked her to start at the beginning.

Tara was a primary-school teacher. Like Karin, she was used

to good health and was always busy. Her work was physically demanding, requiring her to spend most of the day on her feet, often carrying piles of supplies around. She sat on the ground to play with the children and frequently had to work at awkward heights.

'When I decided to be a teacher, I had no idea what a physically gruelling job it could be,' she said. Tara often had mild backache, which she attributed to the strain on her muscles of the work she did. Then, one day, she told me, 'I bent forward to talk to one of the children, and pain like lightning spread from my lower back into my left leg.' The pain was so severe, she thought she would collapse. She had to sit on the ground, while a teaching assistant went for help.

The school nurse came to help her walk to an office, where she took some painkillers and lay on the floor. The pain eased slightly, but she couldn't return to class and was off work for the next few days. Sitting was so painful she couldn't bear it, while walking made her feel fragile, as if the pain would come shooting back at any moment. As a result, she spent the whole time off work lying down, with a pillow under her knees.

She saw her local doctor, who advised her that she had sciatica, and she was referred to a physiotherapist as a first measure. She found the recommended exercises too painful to do. When she developed a burning sensation in her left leg, the physiotherapist referred her to an orthopaedic surgeon for an assessment. Ultimately, a scan showed that she had a minor slipped disc in her lower back. It was pressing very slightly on a nerve, but the surgeon didn't feel that it required an operation and suggested she continue with physiotherapy. Because Tara had already tried that and it had failed, she was left feeling frustrated by the suggestion. She sought a second opinion and got the same advice.

'Every time I moved, I could feel the disc moving. I knew it was getting worse, but it was as if the doctors were ignoring how I felt. I had visions of it cutting right through my spinal cord. I said to one doctor, "Is that what has to happen before somebody will do something?"'

She was, by this time, in constant pain. The burning had moved upwards to involve her whole left leg, as far as her waist. Even stronger painkillers gave her no relief, so one surgeon prescribed painkilling injections. They helped, but so briefly they had to be repeated. By now, several months had passed and she had only managed a handful of days at work. She walked with a limp. She needed sedating painkillers to help her sleep.

As the months rolled on, Tara noticed new symptoms. Numbness started to creep up her left leg.

'It was burning on the inside, but the skin was numb,' she explained. 'You could stick a fork in me and I wouldn't feel it.'

Over time, the pain and numbness moved to the right leg. She began to struggle to walk and her legs were clumsy and weak. Tara was certain the disc had moved and insisted on another scan. When the scan showed no change, she was referred for further physiotherapy.

'I asked them to take the disc out, but they wouldn't,' she told me. 'I couldn't understand how they could just watch me becoming paralysed and do nothing about it. I used to run 10K regularly, and within a month I was walking with one stick, then two sticks, and in the end I had to get a wheelchair. They wouldn't even give me that; my father had to buy my first wheelchair for me.'

Tara had a chronic dull pain, with intermittent waves of deadening pain. During one of these, she collapsed. She was walking around her own home, using furniture to support herself, when pain ripped from her back, down her legs. She blacked out on the spot. When she woke up, her left arm was partially paralysed.

I met Tara some time after these events. She was in an electric wheelchair, and her left arm sat motionless in her lap. She controlled the chair with her right arm, her only functioning limb. When she'd reported losing consciousness to her doctor, he'd advised that he wanted to take her wheelchair away.

'He said that the chair wasn't safe for a person who experienced blackouts,' she told me. 'My hand could get stuck on the controls and I'd end up in the middle of the road. But I only fainted once, and, if I didn't have the chair, I'd be trapped at home.'

I asked to examine her and found that the weakness and sensory loss were entirely consistent with a functional neurological disorder. Nerve, muscle, spine and brain damage cause particular patterns of neurological deficits, which depend on the very complex anatomical arrangement of the nervous system. For example, a lesion in one part of the spine will affect the nerve pathways for joint position and thus affect balance only, whereas a lesion in a different section could leave balance relatively unaffected and cause numbness instead. Damage to one side of the brain causes symptoms on the opposite side, and so on. The distribution of Tara's sensory loss and muscle weakness was incompatible with anatomy. What's more, there was a huge disparity between those clinical signs affected by conscious control, such as strength, compared with those that are unconscious, such as reflexes. Tara's reflexes were entirely normal, even in a limb she couldn't move at all. The diagnosis of FND is made on positive signs of discrepancies found on examination, ones that make a pathological process impossible. The slipped disc seen on Tara's scan was very low in her back, nowhere near her spinal cord, so it could not possibly have caused paralysis in both her legs, never mind her arm. I assumed her previous doctors had thought the same, which is why they had treated her conservatively and repeatedly advised physiotherapy.

When telling me her story, Tara did not let me know what the other doctors had said was causing her almost total paralysis. A young woman losing the power in her legs is a medical emergency and she had been thoroughly investigated, so at least one of the doctors must have given her a diagnosis. I asked her what that was.

'I have a slipped disc,' she repeated, as if I hadn't understood.

'Did anybody say this could be a functional neurological disorder?' I asked.

No, she had never heard that diagnosis before, she told me.

'Did anybody say this could be a psychosomatic problem?'

'Somebody said it was psychological, but that just meant they didn't know what was wrong! How could stress cause this?!' she said, looking utterly astonished that somebody could have said that to her.

I understood Tara's bewilderment. Her disability, which had started with musculoskeletal back pain and a slipped disc, was being described as a psychological disorder – that would baffle anyone. The confusing way that terms like 'psychosomatic' and 'psychological' are used, and the artificial division between physical and psychological ailments, threatened to push Tara into a frightening realm in which people would try to find Freudian subtext in her paralyzed leg.

Unlike Karin, her disability didn't come from stress. Tara's functional disorder had started as a purely physical pathology: a pulled muscle and a slipped disc. That led to evolving physical consequences, but also had psychological consequences.

When the body is healthy, we take it for granted. We can function very effectively without giving it a thought. Throughout any given day, the body undergoes multiple changes, which the brain assesses as normal and duly dismisses. Occasional palpitations walking up a flight of stairs; a small ache in the lower back, caused by sitting in an uncomfortable chair; altered bowel habit triggered by diet; dizziness on standing suddenly – these are only a few of the huge array of small bodily sensations that happen in varying degrees every day. They create a constant background of unobtrusive bodily white noise, to which we rarely give a second thought – unless, that is, something happens to make us pay attention.

If a person has always been well and has no expectation of ill health, they will barely notice how the body reacts to activity and the environment. Their attitude to bodily changes might be different, however, if they are given good reason to notice. A person with a relative who was recently diagnosed with a serious heart condition might find it hard to dismiss palpitations that would normally go without remark. Somebody recovering from cancer might worry that fatigue could be a sign of something sinister.

The body offers an ever-present potential symptom pool. There are numerous reasons why someone might start to pay undue attention to their body and, out of the white noise, pull one sensation to the fore, starting a medical hunt. Once you assess a bodily change as abnormal, it becomes a symptom. A person with back pain or a slipped disc would have reason to pay more attention to their legs than they might ordinarily. A person who believed that a slipped disc was cutting through a nerve would reasonably want to assess the sensation in their legs for a sign that there was nerve damage. When you seek, you find.

As I've already said, the processing of sensory stimuli is subject to many unconscious controls. One I have not yet given the attention it is due is the concept of filtering. At any point in time, only a fraction of potential sensory experiences available to us are in the conscious realm. Where I am sitting now, there are children playing outside, but I have been blocking out their chatter so they won't distract me. I didn't feel the chair pressing into my skin until I started to think about it. I have a slight ache in my left arm from a minor injury, but I forgot about it until just now. I was unaware of how bad my posture had become as I sat hunched over my desk, but, now I am aware of it, I am trying to sit up straighter. There are so many bodily changes and sensory experiences available to us that we would not be able to concentrate if we had to think of every one, all the time. As a result, the brain filters out the excess. Beyond choosing where we want to focus our attention, most of this process occurs at an unconscious level. We have significantly less control over our senses than we think.

Filtering, control of posture and movement are all changed by the attention we pay to them. On a day-to-day basis, we don't give much thought to what our legs feel like or how they are moving or are positioned. The knowledge that she had a slipped disc gave Tara a reason to worry about her left leg, so she began to pay attention to it. As soon as she did so, she noticed every tingle and ache that her brain would usually filter out. That made her concerned, and the concern made her even more attentive to her left leg. She didn't know much about the anatomy of the spine, but became concerned that a slipped disc could slip further, so she began to search other parts of her body for evidence of this happening, and started to notice that her legs felt strange.

Her movement was affected next. We take complex muscle activity, like walking, for granted, because we do it automatically. Millennia of evolution have made it natural for a healthy infant to learn to stand and walk, and for adults to forget what a sophisticated process this is. It is easier to program a computer to beat a chess grandmaster than it is to create a machine that perfectly mimics the human gait. Losing the automatic nature and unconscious control of walking can make the process less efficient. For example, I could walk along the top of a low wall without difficulty, but ask me to do the same on a very high wall and I would become so aware of my balance and movements that I would risk falling. Thinking about movements affects their quality. Tara's concern that the disc would affect her legs caused her to pay more and more attention to her movements, until they started to feel unnaturally awkward. Hypervigilance disturbed the natural quality of her gait. Complex motor activities reside in our procedural memory (muscle memory), to be used without thinking, like riding a bike. But such memories can be unlearned.

The more I talked to Tara, the more I understood how the vicious cycle had unfolded. There were multiple contributors to her disability. The pain changed the way she walked, but – worse than that – it also changed the way she thought about her body. It made her inactive, which deconditioned her muscles. Unaccustomed to being ill, she was frightened. Learning that she had a slipped disc was a huge shock. It made her feel as if she was getting old. Her mother suffered from arthritis and had been in constant pain for decades. Tara feared that her condition was the start of a similar decline. When somebody referred to the slipped disc as 'unstable', it created a vivid, visual story inside Tara's head. When she moved, she could see the disc move in her mind's eye. She could see her spinal cord, compressed and shrivelling. Tara's self-told story, which started with a pinched nerve and was further inspired by her family history and what she believed she knew about anatomy, was as compelling as the story of grisi siknis or the fear of a poisoned uranium mine. She could visualize the disc inching towards her spinal cord, and the physiological tricks and glitches of the brain did the rest.

I explained the diagnosis to Tara, telling her that, as a result of the pain, she had become hypervigilant to sensations and movement, and in doing so had literally retrained her brain so that she could no longer utilize motor activity effectively. I also explained that it could be reversed. Like learning a sport, she could be taught to walk normally again. She looked doubtful. For the moment, I avoided mentioning that the treatment would require intensive physiotherapy, so as not to find myself associated with previous doctors whom she believed had failed her.

'It's the disc, it has to be,' she countered.

Of course, she was right. The disc had been pivotal to the development of her disability, but not because it was pressing on her spinal cord. It set off a cycle of perception and reaction, and without it none of the rest would have happened. In the absolute broadest sense of the term, her disability did have a psychological cause. In other words, cognitive processes that belong to the mind – attention and perception, in particular – were integral to the development of the problem. Anxiety was also very important, because it directed her attention to her legs. However, it was not psychological in the pejorative or reductionist misinterpretations of the term. To many people, 'psychological' is often taken to imply the presence of significant social stressors, mental fragility or psychiatric illness. That didn't fit Tara's experience, so she rejected the initial diagnosis. What was most important for Tara to understand was that her inability to walk arose from physiological changes in her brain, not from some ethereal form of the mind that Wikipedia might suggest.

It is misleading to call Karin's misshapen heart organic and

Tara's immobile legs non-organic. There is a problem with how all these medical disorders are conceptualized. By referring to grisi siknis, resignation syndrome and Krasnogorsk's sleeping sickness as the embodiment of anxieties, following cultural templates, or as a language of distress, I know I risk being misunderstood; it could cause them to fall into the hazy category of counterfeit illness that arises in Descartes' idea of a non-physical mind, when actually I mean the opposite. They are as real as any other medical disorder; they arise from the fallibility of cognitive processes and have a corresponding physiological change.

An equally worrying extrapolation from the idea that organic is real and non-organic is not real is to say that non-organic therefore must be less serious and less disabling. A common argument from people who are finding it hard to accept the diagnosis of a psychosomatic disorder is to say that symptoms feel too severe to be considered 'psychological'. But there is no correlation between the severity or chronicity of disability and the type of disease process. A person with multiple sclerosis may have minimal symptoms, while a person with a functional neurological disorder could be bed-bound. Dissociative seizures usually last longer, are more frequent and more likely to require hospitalization than the seizures that occur in epilepsy. 'Psychosomatic' and 'functional' do not imply less severe or less disabling – far from it.

During my discussion with Tara about the cause of her paralysis, she said, 'How could I possibly be doing this to myself? Surely, if it was psychological, I would have snapped out of it by now?' Of course, her problem was not self-inflicted. She held another common mistaken belief that psychosomatic disorders are mostly self-limiting and cannot cause sustained disability. Actually, for many people, they are a self-perpetuating phenomenon. Chronicity is common. Tara expected her brain to have a fail-safe that kicked in when the shutdown had gone too far – just as, when you hold your breath, carbon dioxide levels rise until the brainstem detects the emergency and forces you to breathe again. But psychosomatic disorders don't work like that. Some people get better, but other people's symptoms are intensified by the experience of being sick. Social and medical interactions weave into the way the patient thinks about the problem, and can make things worse. These too are physiological processes outside our immediate control.

The psychiatrist Laurence Kirmayer uses the concept of looping to consider how complex illness behaviour can emerge and then become amplified over time. In Kirmayer's words, 'the ways that we narrate our experience influence our interactions with others in the social world and this, in turn, reshapes our experience.'

Feedback loops that exacerbate functional symptoms happen at multiple levels. Some are purely physiological. The autonomic nervous system and the hypothalamic pituitary axis (HPA) are responsive to experience and emotion. The HPA has an inbuilt feedback loop designed to temper its response in the long term. Processes like attentional bias are often self-propagating – the more you notice, the more you pay attention. These are internal feedback loops that exist at a biological level.

Other contributors to the looping, amplifying evolution of functional disorders are environmental – external. These include social interactions with family and medical professionals, the medical tests, treatment offered, the new medical knowledge attained by the patient, social attitudes and practical matters of insurance and disability payments. These external factors feed back into the physiological processes to affect the course of the illness. The more tests Tara had and the less she understood the doctors' explanations, the more confused and worried she became. That only made her search even harder for symptoms. Doctors were inadvertently feeding the cycle.

Fortunately, Karin and Tara broke out of their loops. The help that got them there was different for each of them, reflecting the different chain of events that produced each woman's specific medical problems. Karin required intensive pharmaceutical and physical treatment to keep her alive, in the first instance. Ultimately, her heart muscle recovered and returned to a normal shape. Later, she made social changes to alleviate the external stressors. Psychological support, exercise and changes in lifestyle helped to sustain her improvement, although the horror of realizing the fragility of life has not fully left her. The experience taught Karin to listen to her body more closely and to pay attention when things were getting too much.

Tara had to learn almost the exact opposite – to avoid the harm of listening too much. She eventually saw another physiotherapist, who taught her to walk again. To do so, she had to unlearn bad habits. Distraction was often key. She started to listen to music as she walked, which took her mind off her movement and gave her gait a rhythm that made it feel easier. Her back pain never fully disappeared, but a psychologist supported her to respond differently when she felt it. Cognitive behavioural therapy helped her to break a negative cycle of hypervigilance to physical symptoms.

Functional neurological and psychosomatic disorders are often a manifestation of a maladaptive response to the mistakes made by the human system of perception. The brain can only express itself through the body and can only learn through the body's interaction with its environment. We develop by trying things out and responding, seeing how that works out and then doing it all again based on what was gleaned from the first try. Our brains are so clever, we don't even know what they are up to. The signals that bombard us are so complex and tricky, they are very hard to read. I am more surprised that we get it right so often and continue to learn than I am by the times we get it wrong.

Resignation syndrome is an extreme example of how powerful the looping effect can be. As I write, Flora and Kezia are no better, six years into their illness. It is likely that a far bigger solution is required for them than can be achieved by their parents, doctors or social workers. Writing about resignation syndrome, Laurence Kirmayer suggested that a detailed eco-social analysis of all the interactions between families, communities and healthcare systems is required if that problem is to be solved.

I worry about Flora and Kezia, because I am not sure it's possible for the physiological changes their bodies have undergone to be reversed after so much time. Children with resignation syndrome do wake up. They regain full motor function over time, although there are no studies to confirm that their psychological recovery is equally complete. Certainly, I agree that an unconditioned body can be trained back to normal levels of physical activity, but what about the brain? The brain is shaped by experience. It doesn't fully mature until a person is in their twenties. Throughout childhood, it is at its most malleable, and the world takes the opportunity to imprint itself. All the games that children play, the risks they take and their social interactions create connections in neural networks that hold knowledge and emotional maturity that will be vital in adult life. Flora and Kezia are in significant danger of reaching adulthood without the benefit of those formative experiences. The artificial division of medical conditions into organic and psychological has made it surprisingly easy for society and the medical system to leave them without active treatment while debating who is responsible for them. They are passive recipients of information, and I suspect their best hope is for somebody to tell them a new story, give them a new narrative to embody, which will break the cycle that has trapped them.

5

Horses Not Zebras

Authority: *An accepted source of information and advice.*

It was 9 August 2017 when CBS News broke the story. A group of US State Department employees, all stationed in Cuba, had developed a serious, but still unexplained, medical problem. The news report was vague but ominous – something was afoot, but they didn't say exactly what. On 10 August, CNN was more emphatic, declaring a suspected targeted attack on US diplomatic staff in Havana, while NBC announced: *US Pulling Embassy Staff From Cuba in Wake of Mystery 'Attacks'*.

It had started in December 2016 when a cluster of American diplomats in Cuba fell ill with a similar array of symptoms: headaches, earache, hearing impairment, dizziness, tinnitus, unsteadiness, visual disturbance, memory problems, difficulty concentrating and fatigue. Within six months, there were seventeen cases. The embassy medical department wasn't able to explain it, and many of those affected were evacuated to the US. There, two teams of doctors – one led by ear, nose and throat specialist and blast-injury specialist, Michael Hoffer, and the other by a neurosurgeon and concussion expert, Douglas H. Smith – carried out a battery of tests. Both teams, working at different institutions, came to a similar conclusion. This was a unique constellation of symptoms that represented a new syndrome they had never encountered before. The problem was referred to as a 'complex brain network disorder' consistent with a 'traumatic brain injury' – but without any history of brain trauma.

What was the cause? The cases had to be related. Aside from the clear link that all the victims were US or Canadian State Department employees stationed in Cuba, they had one other shared experience: nearly all the victims had reported hearing a strange noise moments before their symptoms started. Those who subsequently described it offered a wide variety of sounds – a grinding noise, loud ringing or high-pitched chirping like a cicada. One said it felt directed and it followed them around their home. Others said it was like a wall of noise when heard in one spot, but it could be escaped by moving position. One person woke in the middle of the night with the sound of loud ringing in their ears. With nothing to go on but the victims' vivid accounts, speculation began that they had been subjected to some sort of sound energy or sonic attack. Intelligence agencies and medical experts involved in the investigation agreed such an attack was largely unprecedented, but possible. The hunt began. According to press reports, the FBI and CIA scoured the diplomats' homes and hotel rooms for traces of a weapon, but found no evidence of one. The medical experts began to use more sophisticated investigation techniques to see if they could detect the aftermath of a sonic attack. When new victims began to present themselves, first in Cuba and then in China, things became even more urgent.

After news of the suspected attack was made public, it quickly caught the attention of the world's press and kept hold of it for some time. In September 2017, the BBC ran a story headlined, *US reveals details of recent 'sonic attack' on Cuba diplomats,* and, the following January, ABC News told their viewers: *US officials still stumped on mystery illnesses in Cuba, open door to 'viral' or 'ultrasound' cause*. Eventually, the phenomenon was given its own classification – Havana syndrome.

Meanwhile, in response to the startling events, US politicians at the most senior level made public statements. Senator Marco Rubio held a hearing, asserting that an attack was a 'given' and that the Cubans were either responsible or knew who was responsible. When asked about the reports by a group of journalists, President Trump added to the melodrama with the half-baked comment, 'some very bad things happened in Cuba'. Meanwhile, senior Cuban officials were denying all knowledge of and involvement in the diplomats' illness.

Experts in the form of doctors, physicists, weapons specialists and engineers now began to offer their analysis of the situation in the press. There was no single consensus, but definite themes emerged. First, no sonic weapon of the type required for this attack was even known to exist. Second, and even more problematic to the sonic-weapon theory, sound is not known to damage the brain, so it was difficult to associate the sound heard by the diplomats with the brain injury described by their doctors. A diagnosis of mass hysteria was put forward early in the discussion, with the head of the human motor control section of the US National Institute of Neurological Disorders and Stroke (NINDS), Mark Hallett, commenting in the *Guardian* newspaper, 'From an objective point of view it's more like mass hysteria than anything else.'

The full clinical manifestations of Havana syndrome, with test results and patient characteristics, became generally available in February 2018 when a paper describing the features of twenty-one victims was published in the *Journal of the American Medical Association* (*JAMA*). Until then, external experts had only had the news reports to go on. The *JAMA* paper concluded that 'these individuals appeared to have sustained injury to widespread brain networks without an associated history of head trauma.' The paper directly confronted the issue of mass hysteria, saying, 'neurological examination and cognitive testing did not reveal any evidence of malingering . . . Rather than seeking time away from the workplace, the patients were largely determined to continue to work or return to full duty, even when encouraged by health professionals to take sick leave.'

Despite the paper's emphatic conclusion pointing to a brain injury, none of the tests described in the *JAMA* paper actually proved that: the brain scans were all normal. Nothing in the paper even confirmed the existence of a brain disease, though that is not necessarily unusual – many neurological diagnoses are speculative and rely mostly on clinical findings. In actively dismissing a functional or psychosomatic explanation, Dr Smith, one of the lead authors of the *JAMA* paper, remarked that, before he saw the patients, he was very dubious about the whole story. However, after meeting them, he realized that, 'there was not one individual on the team who was not convinced that this was a real thing.' He went on to say, 'To artificially display all of these symptoms, you'd have to actually go and research, practise, be the most consummate actor ever, and convince one expert after another.' Still, the lack of proof of and disbelief in a sonic injury encouraged external medical experts in various fields to assert more strongly that this was a case of mass hysteria – or mass psychogenic illness (MPI), to use its more modern name.

The debate raged on, bringing with it some very serious political consequences. The US removed half its existing staff from the embassy in Cuba, leaving only a skeleton service.

The visa processing office was particularly affected, which in turn impacted on travel. A recent explosion in tourism from the US to Cuba receded and the US began to expel Cuban diplomats from Washington DC. Cuban experts defended the country with complete denial, and delicate relationships began to sour. If things were to improve, somebody needed either to find the weapon or find proof of a diagnosis, any diagnosis.

As the back and forth continued between politicians and in the press, Dr Smith's team kept up their search for objective evidence. Nearly eighteen months later, they published their second paper in *JAMA*, which described more advanced neuro-imaging findings in a subset of the diplomats. Compared with a control group, they appeared to have reduced brain volume. Was this proof? The authors didn't suggest that it was, only concluding that further study was required. It was another warped piece of an incomplete jigsaw. But it did lead to a further glut of headlines in the newspapers, each of which cherry-picked from the conclusion to best support its own view. Some emphasized the apparent abnormalities on the scans: *Brain scans hint the mysterious 'sonic attack' in Cuba was real*, while others were more sceptical about the significance of the results: *Don't put on your tinfoil hat yet.*

I cannot say that I know for certain what caused Havana syndrome, although it will not come as a surprise to learn that I agree strongly with the experts who said this was more like a functional neurological disorder or mass psychogenic illness (MPI) than anything else. However, even without having proof of a diagnosis, I think there is a great deal to be learned from looking at the events in Havana. It seems to me that the incident has trapped a group of people in a web of politics, commerce and, perhaps above all else, pride. Some of those people are still in that trap.

Let's start with the sonic weapon, which has a pivotal role in driving this story forward. From an early stage, weapons experts were very clear that a weapon of the sort suspected to be responsible for the attack was not known to exist. Medical experts were equally clear that sound did not cause brain damage. In fact, Dr Smith's team acknowledged clearly in their first *JAMA* publication their struggle to associate the sound with the suggested brain network disorder. In the body of the text, for those who cared to read the paper in detail, the authors stated, 'sound in the audible range is not known to cause persistent injury to the central nervous system.'

If everybody, including the doctors central to the story, agreed on the implausibility of the sound-weapon theory, how did it remain a feature of the investigation for so long? I would suggest it took hold as false news generally does: it was politically convenient, simple and believable to the non-expert. It fitted neatly into a political climate of long-standing suspicion between two nations. It was also exciting and controversial. The spy craft was thrilling for the media, providing great fodder for headlines. All of those involved in the investigation must have felt they were at the cutting edge of a conspiracy. I would also suggest this feeling was so compelling, they struggled to let go and risk thudding back to normality.

The community's ability to disregard the inconvenient truths that sonic weapons don't exist and sound doesn't damage the brain was also an example of a common response to the experience of cognitive dissonance. This refers to the discomfort we feel when faced with information that doesn't match a strong belief. Such is the feeling of unease created by cognitive dissonance, it often sees us rationalize what may at first seem like an irrational opinion or choice.

False beliefs, like the certainty of the existence of a sonic weapon, are at the heart of the development of many functional disorders. They create the expectations of illness that code themselves into the brain. Tara carried the false belief that her slipped disc could sever her spinal cord, even though the position of the disc meant that was actually impossible. But, for Tara, this belief was so strong, she found ways to dismiss expert opinions. She sought out any bit of information that would support her view, and gave it prominence. In the same way, impressive leaps of imagination were needed to keep the sonic-weapon theory alive.

The sonic-weapon narrative appeared to arise from the accounts of the index case – patient zero. It's important to remember that the index case in an outbreak like this often has a different medical problem from those who follow. But they set the scene. They create the template from which future victims unconsciously draw their symptoms. I can't know for sure what caused patient zero's acute onset of dizziness, hearing impairment, tinnitus, loss of balance and fatigue, but they did associate their symptoms with hearing a sound. From this, three possible scenarios can be postulated.

First, patient zero was actually attacked by a noxious agent and, whatever form the attack took, there was an associated sound. Attacks on agents and politicians are well attested, and, while the sound could not have been the medium for damage to the nervous system, it could have been a by-product of some other sort of attack. I don't favour this explanation, but it is entirely within the bounds of possibility.

Second, patient zero could have developed an illness for any of a dozen other reasons, and, in searching for an explanation, they remembered having heard an odd sound. Our brains hate chaos; we always want to know why something happened, and it's human nature to search recent experiences for that explanation. People with a newly diagnosed disease often hit on the memory of a recent minor injury or some kind of environmental exposure on which to pin the blame, when, really, many diseases occur just by chance. A person will hear a lot of unexplained sounds in their lifetime, but they are more likely to recall the odd sound they heard shortly before they developed a significant illness. This is called recall bias, in which small things that happen just before a major life event take on much greater significance than they deserve. In this scenario, the sound was a coincidence that took on prominence through recall bias.

The third possibility is that patient zero's illness was functional from the start, created by the anxiety of a suspected attack. If that person was in a position to think they were at risk of attack, then hearing a sudden sound might have worried them enough to provoke them to search their body for injury. Bodily changes are always available to notice, if we look for them – it's that white noise again, creating a constant pool of physical symptoms which we usually disregard. A sound would draw attention to the ears, making the head the focus of the symptom pool.

To the outbreak at large, it doesn't matter whether patient zero was right or wrong to associate hearing a noise with the onset of their symptoms. Nor does it matter if the sound or the illness came first. What matters is that key people believed the sound was associated with an attack. According to media accounts, patient zero was trained in and sensitized to covert surveillance, so they must have made a compelling and believable witness. That person, or somebody to whom they told their story, came up with the idea of the weapon, and the early believers were powerful enough to drive the conspiracy theory onwards. The idea of an embassy under attack fits well with decades of suspicion between the US and Cuba. On the surface, the sound weapon was feasible – at first glance, at least. So potent were those early proponents and so apparently credible was the idea, it allowed swathes of people to subsequently bypass the flaws in the concept. Sustaining the sonic-weapon theory required numerous illogical leaps that ignored the absolute implausibility of the whole scenario.

Functional neurological disorders are usually anatomically and biologically impossible, a feature that is often central to making the diagnosis. The symptoms come from the unconscious and are based on people's understanding of how the body works – but those understandings are usually inaccurate. It is easy to see how people came to believe that sound could cause brain damage. Sound enters through the ear, which creates the impression that the ear acts as a conduit through which sound has direct access to the brain. The sonic-weapon story presented a very attractive mental picture in which a noise entered the head via the ear and, from there, shocked the brain. But that is an anatomical nonsense. The ears are not a direct conduit to the brain; they are sensory organs, like the skin. Sound stimulates the tympanic membrane, it vibrates and ultimately the sound is converted into an electrical signal that travels along a nerve to the brain. All sensory signals are transmitted by nerves, and hearing is no different. A wave of sound energy would have no more direct access to the brain than it would to any other organ. A burst of energy sufficient to damage the brain would damage other organs, too. The idea that sound causes preferential brain damage is therefore a sort of folk illness based on false beliefs about anatomy and physiology.

Of course, loud sounds can cause hearing loss by damaging

hair cells within the ear. It's feasible that a person exposed to a very loud sound could develop hearing loss and dizziness, although not the 'brain network disorder' postulated by the medical experts involved. But, to believe a sonic weapon had damaged the ears, with the brain injury being a red herring, requires us to gloss over another set of inconvenient facts. Most noteworthy are the circumstances of the 'attacks'. The victims were said to be targeted in multiple locations – in crowded hotels, in private villas, on the twentieth floor of apartment blocks – seemingly in any part of the city. In one example, a doctor sent to Cuba to work for the State Department heard noises almost as soon as he checked into his room at the Hotel Capri. His visit to Havana had not been announced, but, if his story was to be believed, the attackers were prepared for his arrival. In a comparable case, a CIA agent was said to have been attacked very shortly after checking into the crowded Hotel Nacional. Sound loud enough to damage the ears would be heard by everybody within range, but only the victims of Havana syndrome heard the sound. The weapon needed to perpetrate these attacks would have to be small enough or far enough away not to be noticed. It had to be mobile, focused enough to pick off its victims in a crowd, and it had to penetrate walls. No Cuban nationals, no hotel or embassy staff, no tourists were affected by Havana syndrome. The weapon achieved all of this while remaining undetectable.

What is also fascinating is that, for all the vivid descriptions of the offending noise, nobody had made a recording. Intelligence agencies tried and drew a blank, although there was a thrilling moment when the Associated Press suggested that they had managed to surpass the FBI and CIA and had got the sound on tape. What they had recorded was aired on numerous television stations, and some of the victims were said to have confirmed that it was the offending sound. It was high-pitched and shrill, but apparently lost its potency when recorded, meaning it could be safely played over the airwaves. Expert analysis would later attribute the sound to cicadas.

For all these barriers to the sonic-weapon theory, it persisted for a very long time. It seemed as if so many press conferences had been given asserting the certainty of its existence, it had become untenable for those who held that view to back down. Like all good viral ideas, it was incubated in reality, which was probably what made it feel so viable. The US community in Cuba had reason to be concerned about attacks and covert home invasions. The history of the region was full of stories of foreign spies breaking into the homes of US diplomats just to unnerve them. During the Cold War, United States envoys in Cuba told stories of the power being cut off unexpectedly, or waking in the morning to find fridges unplugged and cigarette butts in ashtrays. Cuban and Russian spies were said to have created paranoia by moving objects around in the home, to let the occupant know they had been there. The United States embassy in Cuba had recently reopened after a fifty-year stalemate, so it would not have been ludicrous for a person working there to feel they could not trust their environment.

There was even some small precedent for mystery energy weapons. It has been said that the Pentagon once tried to develop weaponized infrasound, although without success. Before that, in the 1960s, low-level microwaves were apparently detected in the United States embassy in Moscow. Some speculated that this was an attempt at mind control, others that it was part of an attempted eavesdropping exercise. Sound has certainly been used as a deterrent at borders and on ships and for social control. When used in this way, it can target selected groups. For example, young people can hear high frequencies that older people cannot hear, so high-pitched sound can prevent young people from congregating, without affecting older people. So, focused sound attacks are possible, but there is a big difference between using sound as an irritant to a subgroup of people and using it as a weapon against a single person. In the end, only a small number of people know the sort of weapons the intelligence agencies have on hand. Those unknowns and the allure of spy craft may have helped keep the weapon search alive – but even the assumption of espionage could not change the fact that sound neither damages nor has any particular affinity for the brain. As that fact was slowly acknowledged, it only made way for the next misstep in the investigation, in which those still supporting an attack of some sort moved their attention to the possibility of an energy source outside the range of hearing: ultrasound or infrasound.

Ultrasound is used in medical procedures. It breaks up kidney stones, so using it to damage an organ would not be impossible – but it cannot be fired at a person at long range and has no specificity for the brain, so would not cause preferential brain damage while avoiding other organs. Infrasound was problematic for all the same reasons – and, of course, both were illogical for a much bigger reason. If the entire premise for the existence of a sound weapon was based on the fact that most of the victims had heard a sound, where was the sense in looking for a weapon outside the range of hearing?

To try to get around the 'audible-inaudible device' problem, Dr Smith, the senior author of the *JAMA* papers, among others, suggested microwave energy. He thought microwaves might create air bubbles in arteries, which would create a sound 'in the brain' of the victim that was not heard by others. Quite apart from the fact that there is no evidence for this, how would microwaves selectively affect the brain? Surely, they could create these notional air bubbles in any artery and therefore affect any organ? They would also burn the skin.

Meanwhile, Dr Hoffer, the ENT specialist leading the other team of medical experts investigating the problem, compared the diplomats' 'injuries' to those seen in explosions. Blast injury is caused by a huge wave of pressure, which displaces the person and all their organs. Several of those affected reported feeling 'pressure-like sensory symptoms' in association with the sound. I will not repeat all the reasons this simply doesn't explain Havana syndrome, except to say that a 'feeling' of pressure does not equate to an explosive displacement.

Could there have been another type of weapon or accidental toxin, nothing at all to do with the sound? Perhaps the compelling sonic-weapon theory had distracted from other types of attack. More recently, reports have raised the possibility that an anti-mosquito spray was the actual culprit – but that, of course, does not fit with the reality of a highly selective outbreak, only affecting US and Canadian embassy staff, and picking them off in properties all over the city. The bigger problem was the lack of hard evidence for a disease process. Once again, disease shows itself, even if the cause doesn't – blood-test abnormalities, brain-scan abnormalities. You may be thinking that the second *JAMA* paper reported precisely that – apparent changes in the brain scans of the victim group versus the control group on advanced neuroimaging. Could this be evidence of an attack?

It's important to note that modern brain scans are so sophisticated that they often show irregularities, the clinical significance of which is poorly understood. Imaging techniques like MRI are very new and we have only very recently started to learn what the inside of a healthy body looks like. In the case of the US diplomats, the brain scans were actually reported as normal. What were ultimately reported as potential abnormalities in the second *JAMA* paper were not so much 'abnormalities', but rather differences between the scans of the diplomats and those of the control group. A difference is not evidence for injury and could have an innocent explanation. The authors of the paper were also quite clear that they did not know what the 'differences' meant to the patients – if they meant anything at all.

In the *JAMA* paper, the MRI scans of the affected group were compared with those of a control group. Some participants in that control group were matched to the victims in terms of education and professional status, but many were not. Since the two groups were not closely matched, and therefore probably lived quite different lives, a difference in their brains should not be over-interpreted. Diplomats resident in Cuba almost certainly have quite a different lifestyle to a random selection of people living in the US. They probably engage in more long-haul travel and it would not be surprising to learn they drink more alcohol; I'd also venture that they might even smoke more cigars. What's more, in choosing the control group, the researchers excluded anybody who had a history of any neurological condition or brain injury. There was no such elimination from the victim group. That the brain scans of the two groups had small differences could be entirely accounted for by the study design.

The two teams of doctors, under Dr Hoffer and Dr Smith, held fast for a long time to the likelihood of an attack. In the face of many external experts insisting on a mass psychogenic illness hypothesis, they dug in to support their original diagnosis of brain injury, despite its obvious flaws. There must have been considerable professional pride involved, given the public nature of the case. Of course, it is important to acknowledge that those US specialists were the only doctors who had actually met the patients – no one else even knew their names, since they were protected not only by medical confidentiality, but by their diplomatic status. One could rightly say that the opinions of Dr Hoffer and Dr Smith were the most fully informed and should therefore take precedence; after all, they took precedence in the victims' lives, determining the medical care they got. Once again, the point is not to say whose opinion is correct, but rather to give fair consideration to alternative explanations, ones that do not require so many leaps of faith and jumps of logic.

Doctors' opinions differ all the time, and sometimes we attach ourselves to a line of thought and have difficulty giving it up. I have certainly been guilty of that many times. It is worth giving a moment's thought to the possibility that, for the US teams involved in this case, the adventure of finding themselves part of a search for a sonic-weapon attack drew their attention away from ordinary medicine. As is often heard said in medical circles, 'When you hear hooves, think horses not zebras.' Functional disorders are common. Sonic weapons are not. How was an everyday diagnosis sidelined for one that many thought impossible?

The usual way of making a medical diagnosis is to take the clinical signs and symptoms, and use them as signposts to indicate which organ is diseased. Having located the problem in the body, the next step is to wonder about the underlying pathology. But, in describing the typical features of Havana syndrome, the *JAMA* paper took the reverse approach. The stated purpose of the paper was 'to describe preliminary findings from 21 patients who were exposed to the same non-natural source'. The attack was presented as an assumed truth. The diagnosis was in place and it was up to the authors to retrofit their findings to the weapon.

The *JAMA* paper concluded that the diplomats had a 'brain network disorder'. But what does that mean? Truthfully, it doesn't really mean much at all. It is uncertainty masked by technical terminology. It's a description of a set of symptoms that confirms the view that the people in question were indeed genuinely unwell. It is meaningless in terms of a specific medical diagnosis. Functional neurological disorders arise as a result of changes in the connections between brain regions, so they would fit neatly into the general category of a 'brain network disorder'. So would head injury and numerous other medical problems.

Dizziness, headaches, tinnitus, memory impairment, unsteadiness – these are some of the most common medical complaints there are. Almost every family doctor, neurologist and ear, nose and throat doctor hears these exact problems described several times a week. So why, when the external medical community was proposing the FND or MPI diagnosis (in a variety of guises), was it so summarily dismissed by those involved in the case? Of course, the teams who met the patients may have had information we do not. (Although, surely, if there was a key piece of information, it would have been included in one of the academic publications, and there was nothing convincing there.) I fear their statements to the press suggest the real reason for the rejection of a psychosomatic explanation. One said that what he had seen was not 'just hysteria' and what he had measured could not be 'faked'. The *JAMA* paper dismissed a functional disorder by saying the patients weren't malingering – as if the two were the same, when they absolutely are not. Unlike functional disorders, malingering is not a medical problem; it is fraudulently and deliberately pretending to be sick. The diplomats' keenness to return to work was also used as a defence against the MPI diagnosis, a statement which implies the doctors in this case held the mistaken belief that people with functional disorders do not want to work. In his press statement, the lead *JAMA* author even equated psychosomatic disorders to acting. The evidence suggests that some doctors still do not know the difference between malingering and psychosomatic and functional disorders. If you do not believe that functional illness is real, then of course you will defend genuinely suffering patients against it. Sadly, however, it is this exact conflation between functional and faked that explains why many ordinary people are devastated by the FND diagnosis – they fear the judgement it will bring. What would these doctors say to Kezia and Flora, who have been bed-bound for five years? Or to Tara, who couldn't walk or care for herself? This view of functional disorders could have pushed the Havana victims into a corner. Either they believed they were under attack, or they were not 'really' ill. There was a covert weapon, or they were faking.

It is important to consider the consequences of a realistic diagnosis being neglected in favour of a fanciful, far-fetched sonic weapon. If this was the case, how much time passed and what might have been lost by leaving a treatable, ordinary medical disorder untreated?

Mass psychogenic illness tends to happen in contained communities under strain. The circumstances for a mass illness to occur were rife in the US embassy in Havana in 2017. But, if the symptoms were indeed functional or psychosomatic, as I attest, then where did they come from? They were not conjured from thin air. How did so many previously healthy people all develop very similar symptoms in such a short space of time? As was the case for the people of Krasnogorsk and the Swedish refugee children, one assumes that embodiment and predictive coding caused the diplomats to play out a narrative written by others. Like Tara, when prompted to be concerned about their health, they examined their bodies for signs of disease. But there are other ways of thinking about how new syndromes evolve. For example, the work of the philosopher Ian Hacking offers a very enlightening viewpoint on the subject.

Hacking described phenomena which he called 'making up people' and 'looping' (earlier discussed, in a different context, in the work of psychiatrist Laurence Kirmayer). 'Making up people' refers to the way in which new scientific classifications bring into being a new kind of person. In other words, once you give a person a label, that person is encouraged to take on the features of that label – this is also called the classification effect. The change occurs at an unconscious level and it is reciprocal. When that new person joins the classification, they bring their unique self to it – and, in turn, change it. This is the looping effect – the classification changes the person, who in turn changes the features of the classification.

Hacking used multiple personality disorder to illustrate the effect of classification on people. Psychiatrists began making the diagnosis of multiple personality disorder in earnest in the 1970s. Before then, descriptions of split personalities were exceptionally rare. Once they started to emerge, 'unhappy people' (as Hacking referred to them) found something familiar in the description and took on the label to explain their own difficulties. Others had the label given to them by doctors keen to offer a diagnosis. (Doctors and patients can only use the diagnostic classifications available to them at a given point in time, and sometimes have to wait for the right one to come along.) As soon as people had this new label applied to them it was inevitable they would search themselves for other symptoms and signs known to be associated with the disorder. Through that process they inadvertently took on features of the label and thus become changed by it. This is an example of Hacking's new kind of people – 'making up people'.

At the beginning of the 1970s, the initial new cases that emerged of multiple personality disorder were relatively simple and ordinary, with people experiencing two or three dissociated personalities, none of which were necessarily extreme. These cases were somewhat similar to the sporadic case descriptions of the century before. But as new people were given the label, inhabiting it and bringing their own selves to it, the symptoms evolved, becoming more bizarre. New, more complex cases, with large collections of extreme personalities emerged. Clinical criteria then changed to incorporate the new manifestations of the newly affected subjects – the looping effect, in other words.

Remembering that the mind extends into the environment, as neatly demonstrated by David Chalmers, we can recognize that external factors were important in the evolution of multiple personality disorder. It was not just an internal psychological process belonging to individuals. Not least of those external stimulators was the media, which took to this sensationalist subject with gusto. Undoubtedly important was the 1973 publication of the book *Sybil* and the very successful 1976 television movie that it inspired. Sybil was a young woman whose traumatic upbringing was said to have caused her to develop seventeen distinct personalities. These were mostly children, male and female, and ranged from the innocent to the death obsessed. Sybil was compellingly portrayed by the actress Sally Field, who won an Emmy for her performance. The film's success ensured that the diagnosis of multiple personality disorder entered public consciousness.

Medical professionals were also important to the story of the disorder. They validated the diagnosis and recruited new members to the classification, forging the necessary alliance between patients and doctors that would create the phenomenon.

A person who was told they were under attack would have a very good reason to scan their body for signs of ill health. Knowledge of the sorts of symptoms expected would focus that search on particular parts of the body. As previously detailed, the body is awash with white noise, so symptoms can always be found if a person looks hard enough. The autonomic arousal of fear would heighten the white noise. Once the classification that was 'Havana syndrome' became available in the United States embassy, it could be used (unconsciously) by frightened people to explain how they felt. It could also be used by doctors to offer a diagnosis to distressed people. Hacking's ideas take the spread of functional disorders a step further by reminding us that each new person has the potential to bring something of themselves to the new diagnosis, and that creates an evolving, rather than a stable, symptom pool.

Hacking also wrote about resignation syndrome, suggesting that early cases entered the social conversation through widespread media coverage. Refugee groups, while from disparate places, shared a predicament and an environment within Sweden. This led to imitation of the symptoms and eventually to the unconscious embodiment of those symptoms. In other words, vivid models of symptoms became salient within the children's environment until they were owned and enacted.

Laurence Kirmayer applied Hacking's classification effect to resignation syndrome: naming it legitimized it as a medical problem and created a classification through which the suffering of future immigrant children could be explained. Social factors, such as growing hostility to immigration in a country that previously prided itself on being receptive to asylum seekers, were undoubtedly important. Societal pressure created a medical drive to explain the disorder in a biological way, to mitigate against the social implications. The classification created a new behaviour in doctors as well as patients. The children became the new kind of people; classification influenced medical professionals' responses to the children. Media reports, politicians, concerned doctors and citizens created a narrative around the new illness classification, new cases were reported and the looping effect played on.

In the case of the people of Krasnogorsk, once the idea of environmental poisoning had been postulated, worried people looked for the typical signs of illness. Some noticed symptoms that they thought could be related to the classification. The presence of a sleeping sickness in their town 'made new people' – those who took on or were given the classification. 'Unhappy people' (as Hacking called them) look for explanations for their problems, and, when the sleeping sickness came along, they duly attached to it. The sleeping sickness became a diagnosis newly available to doctors, so they could use it to explain symptoms that might previously have gone unexplained, but every new victim brought a personal element to the symptoms and something specific to their experience of it. As I was constantly reminded by the people I met in Krasnogorsk, different people got the illness differently. Some acted on autopilot. Some slept. Children hallucinated and plucked at their clothes. The typical features of the poisoning changed over time through the process of looping. New symptoms invited new members into the classification and the process continued. Then the media arrived and added to the phenomenon, while researchers spent months examining the atmosphere for toxins, amplifying the cause for concern. And so on.

The concern raised by Hacking's work is that, as new types of people emerge out of new classifications, people become moving targets to psychiatric and behavioural scientists, as is well demonstrated by the ceaseless renaming and philosophical rebranding of psychosomatic disorders. These problems are constantly being reimagined by new generations of doctors and patients. Charcot's hysteria caused by brain lesions became Freud's psychogenic conversion disorder, which, in turn, after a few other incarnations, became functional neurological disorders. But none of the new classifications or reimaginings have really solved the problem. They all seem to be too reductionist, with each working very hard to leave behind at least one of the three elements of biopsychosocial illness.

Like resignation syndrome and Kazakhstan's sleeping sickness, Havana syndrome is not about the vulnerability of the individual; it is about the sociopolitical environment in which the story unfolded. The classification had momentum because it proved to have benefits for the doctors who had created it and for others who had a vested interest in Cuba–USA relations. It also exteriorized tensions within the embassy, in the same way that grisi siknis was used to exteriorize conflict in the Miskito community.

The groundwork for Havana syndrome was probably laid down long before those involved had even considered making Havana their home, before the doctors were doctors and before some of those involved were even born.

I was not alive during the Cuban Missile Crisis, but it was a sufficiently significant event for it to be well known, even by those of us who grew up thousands of miles away, years after it happened. Those closer to the story – the US embassy staff, for example – must have had a great deal of history and suspicion to overcome when the US embassy in Havana reopened in 2015, after more than fifty years of trade embargoes and hostility.

Diplomatic ties between the US and Cuba were severed in 1961. The following year, a thirteen-day stand-off brought the US and the Soviet Union as close to nuclear war as they ever got. Russian missiles positioned in Cuba were only a short flight from the US mainland, making the threat very real. The crisis was averted, but relations between the US and Cuba remained cold, and travel and financial transactions were banned between the two countries for decades. Then, in December 2014, President Barack Obama and Raúl Castro announced a softening of relations between the two countries, opening up the possibilities of trade and tourism. A year later, they reopened their embassies in the respective countries.

US diplomatic staff moved to their new jobs and their new homes in Havana in 2015. The building that houses the United States embassy in Havana is a glass and concrete tower, which must have stood out when it was first built in 1953. It had lain half dormant for fifty-four years, under the protection of the Swiss government. That is not to say that it lacked for excitement during the years in hibernation. In 1964, the Swiss ambassador had to resist the Cuban government's attempt to seize the building, and, in 1970, when a Cuban fishing boat was captured by Cuban exiles, the Cuban public laid siege to the building for three days. The abandoned embassy finally got some new residents in 1977, when, following an agreement between Jimmy Carter and Fidel Castro, a small group of US diplomats moved into the building, although it remained under Swiss jurisdiction and the flying of the US flag was prohibited.

The years of neglect had not been kind to the building. Its position on the esplanade, facing the ocean, meant that saltwater had eaten away at the marble facade and, before it could be reopened, it had to be heavily refurbished. Wary that Cuban spies might install secret listening devices during the renovations, maintenance crews were not allowed above the second floor. Cuban staff were hand-picked and vetted, and supplies and furniture were shipped in from the US. On 20 July 2015, the United States embassy seal was proudly restored and the embassy resumed its role. On 14 August 2015, then Secretary of State, John Kerry, the most senior official to visit the country for seventy years, travelled to Cuba to raise the US flag. Less than a year later, in March 2016, Barack Obama was the first president to visit Cuba since Calvin Coolidge, in 1928.

Negotiations to get to that point had taken years. The move had faced off opposition, particularly from Cuban exiles resident in the US. Senator Marco Rubio, who entered the 2015 race to be the Republican presidential candidate, called it a 'concession to a tyranny'. Fellow presidential hopeful Ted Cruz shared this sentiment, writing in *Time* magazine that 'America is, in effect, writing the check that will allow the Castros to follow Vladimir Putin's playbook of repression.' Suspicions ran high at the formal reopening. The USA feared espionage within the walls of the newly refurbished embassy, not least since diplomats stationed at the embassy while it was under Swiss supervision had reported haranguing tactics by Cuban spies. The Cubans were also suspicious. An influx of US diplomats and their families potentially eased the way for US intelligence agencies to increase their activity in the country. Some Communist Party members, including Fidel Castro himself, made cautionary, anti-imperialist statements after Obama's visit.

As exciting as it must have been to be the first to occupy the old building after half a century, it undoubtedly caused some trepidation, too. Its success was also short-lived. Following

Fidel Castro's death in November 2016, the newly elected US president, Donald Trump, issued a statement threatening to cancel Obama's deal with Cuba as soon as he took up office. In June 2017, he followed through on this threat and rolled back many of the Obama administration's efforts to normalize relations between the two countries. The embassy had barely had time to dust off the cobwebs.

It was in December 2016, a month after Trump's election, in an atmosphere of uncertainty and hostility, that the first person reported hearing strange noises in his home. By February 2017, there were six confirmed cases. Not all said they heard a noise, but most did. With a suitably dark sense of humour, the earliest name given to the alleged attacks and subsequent unexplained symptoms was 'The Thing'. It was around this time that Audrey fell ill.

Audrey (a pseudonym) was an experienced Foreign Service staffer. She told her story to the *New Yorker*, and it provides a compelling example of how mass psychogenic illness spreads. Cuba was a new posting for Audrey, but one that excited her. When the first odd thing happened to her, she didn't give it much thought. She and her family had returned from a holiday to find that their usually airy Spanish-style suburban home smelled of rotting food. On investigating, she found that her freezer had been unplugged. Audrey managed to brush the incident off, but was reminded of it when, a little while later, she fell ill. On 17 March 2017, she was in her kitchen when she felt a sudden burst of pressure in her head. She had never had pain like it. She was aware of the rumours about sonic attacks, but not of the details. The severe headache brought to mind advice that had been given to staff, should they think they were under attack. They were told to move away from the spot where the symptoms started. If the attack was directional, that might alleviate the symptoms. It didn't work for Audrey, and the pain continued into the night and beyond. With time, her symptoms evolved – she became unsteady, had difficulty concentrating and felt dizzy.

Audrey was already symptomatic when, later in March 2017, the embassy's Chief of Mission called a group meeting to enquire after the well-being of the staff and to advise those with symptoms to come forward. Some reported that they believed they had been affected, claiming they had heard noises in their personal apartments on the twenty-first floor of the embassy, in their detached homes and in hotels. Audrey felt unwell, but hadn't heard any strange sounds, so she convinced herself she didn't have The Thing. The Chief of Mission told the group that, in the event that they heard any strange noise in the future, they should try sheltering behind a concrete wall.

Audrey was no better when, in May 2017, the Chief of Mission spoke to staff again. Concern seemed to be escalating, because this time he advised people to have medical checks, even if they felt fine. Dr Hoffer had been flown in to examine them. This time, Audrey was more open about her symptoms – in particular, that she was struggling to balance. After her assessment, she was told that she didn't fit the criteria for the illness and was therefore not a confirmed victim. A month later, she felt worse. It was only at that third presentation that she was added to the list of those who needed treatment. She was flown to Philadelphia and finally had her diagnosis validated.

Between December 2016 and August 2017, the events in Cuba were kept out of the press, but there was considerable internal medical and political response. In February 2017, there were enough cases for it to be considered an outbreak, which is when Dr Hoffer became involved. He was not only an ENT specialist, but was also ex-military, with expertise in blast injury. He flew to Cuba to examine the victims, and those most severely affected were later flown to Miami, where he investigated them further. It was his team that was first to suggest a diagnosis of traumatic brain injury. He found evidence of inner ear damage, but since that couldn't account for all the symptoms, he suggested the embassy staff affected must also have brain damage. The team published their account of the phenomenon in a journal that Hoffer also edited, *Laryngoscope Investigative Otolaryngology*. As with the *JAMA* paper, this account took the 'energy attack' as a truth, with the stated purpose of the academic paper being 'to describe the acute presentation of individuals who experienced neurosensory symptoms after exposure to a unique sound/pressure phenomenon.' While advising caution in coming to any definitive conclusion, and agreeing that the cause of people's symptoms was unknown, the authors also said, 'It would be imprudent to exclude any potential directed or non-directed energy sources at this time.' They went on to speculate on a number of ways a sound weapon could have caused the symptoms and, while acknowledging that their publication served only as a description of the victims' experience, they also stated that the symptoms were very like those caused by 'traumatic brain injury following blast exposure or blunt trauma.' This paper was published two years into the outbreak, after the intelligence agencies had searched for and found no evidence to support an energy weapon. Diagnosis of blast injury requires that a person be in a blast, but even the fact that there was no proof of such a precipitant did not put anybody off the single-minded assumption that that was what had caused the victims' symptoms.

There was an era when every word that doctors said was believed, a time when they were always treated with deference. That has changed, but it doesn't mean they have lost all their power. Within their specialties, individual and small groups of senior doctors can influence the entire course of thinking and research in their area of expertise. They can potentially do so for their entire working lifetime. People embody classifications and classifications are created by doctors. An academic doctor can create belief systems for generations of subsequent doctors, as research tends to follow a path laid out by the most powerful players in the field. A doctor in the public eye can influence health and illness beliefs, and even create health trends among the public.

After I wrote my first book about psychosomatic disorders, people started to contact me with their stories of illness. Having related to something in my patients' stories, they hoped I would give them a formal diagnosis. Often, their descriptions didn't, to my mind, fit in any way with a functional neurological disorder as I had described it. They had identified with something in my description, but then they had added their own personal element to it. I realized I was 'making new people', as Hacking might have put it. When people have health problems they don't understand, they search their environment for an explanation – an explanation generally makes people feel better. Answers might come from television programmes, social media, newspapers, from neighbours and books. Doctors are another source. Many diagnoses are subjective because they are based on a constellation of typical symptoms rather than on a single diagnostic test. That means a doctor can choose to diagnose more cases – or fewer – depending on how lax they are with rules of classification.

Individual doctors have the power to create new medical diagnoses and to influence the course and treatment of illness. I worry about it in my own work. I worry about it for Dr Olssen and the resignation-syndrome children. Sometimes, asylum-seeking families in Sweden go directly to her if they feel their children are becoming withdrawn. They bypass conventional medical routes and ordinary lines of social support, which potentially places the children on the trajectory to resignation syndrome as soon as the contact is made. Dr Olssen is certain that the only cure for resignation syndrome is a positive asylum decision. It's inevitable that she will communicate that idea to the children and their families, even if she doesn't mean to. I heard her say it in front of Helan and Nola more than once. As a result, it is possible that at least some of the children may be embodying her expectations.

I fear the assumption of an attack and the influence of powerful medical specialists may also have influenced the trajectory of Havana syndrome. The affected diplomats were very quickly referred to blast-injury specialist, Dr Hoffer, as if the attack was already a proven fact. They were then sent to a concussion specialist, Dr Smith. Medical efforts took a reverse approach, assuming an energy-weapon attack and subsequently attempting to explain the symptoms through that lens. The people were told they had a blast injury, but without the blast, by the blast-injury specialist, and concussion, without concussion, by the concussion specialist. Scans were scrutinized for any tiny bit of proof that would support these odd contradictory diagnoses. Tenuous 'proof' was found.

The academic papers and public statements from the medical teams involved all toed an awkward line between admitting that they had found no proof of brain disease, while still managing to imply that an attack was a safe bet and that proof of it would come with time. In one interview, Dr Hoffer said, 'the evidence suggests they were targeted, although we can't prove that . . . There could easily be injury in the brain, we just don't know that.'

Havana syndrome was a powerful sociopolitical wave. Many people were caught up in it, including US embassy staff, politicians and doctors. At times, it seemed as though Dr Hoffer could not hide his excitement at being a key player in uncovering international espionage, giving vivid descriptions in interviews of the moment he was called in to consult: the phone rang, he picked it up and someone on the end of the line said, 'Dr Hoffer, this is the State Department, we have a problem.' It must have felt like the start of a Mission Impossible movie, and I don't think this is irrelevant to how things would eventually unfold. One could say that the doctors were caught up in a wave of mass hysteria, as much as any of the people who fell ill.

The staff of the US embassy were caught in a web of powerful forces. Senior doctors said they had brain injury and all but ordered them to search their bodies for symptoms. As anxiety grew, they were advised to seek medical attention even if they felt perfectly well. Healthy people were asked to actively pluck bodily changes out of the white noise and were told to be concerned. Those working in the embassy were instructed by senior staff at the most senior level to be vigilant, to hide behind concrete walls, to pay attention to their bodies. The US Secretary of State summoned dozens of staff back to the US to be examined, and US politicians who had objected when the embassy in Cuba reopened seized on the alleged attacks to prove they had been right all along. Marco Rubio referred to the attack as a 'documented fact', even after the FBI said they had found no evidence for it. High-level politicians and doctors with reputations to protect made some very definite statements in the press. I cannot imagine how hard it would have been to resist developing symptoms in that setting, and how difficult it would have been to accept a mass psychogenic explanation with all the 'experts' disparaging it so. We embody narratives, some of which are embedded in our culture, while others are our own personal narratives. Some are told to us by powerful people – doctors, politicians, activists, public figures, celebrities. What choice did the embassy staff have when the highest authorities were telling them with such certainty that they were under attack?

If a model for illness is vivid enough and the basis for the illness is sufficiently salient, it is easily internalized by the individual and then passed from person to person. It happens because the idea is reasonable and because physiology behaves that way. It's a normal thing that gets out of hand. So potent was the audible-inaudible weapon story that, even after the intelligence agencies ruled it out, the illness spread to China. In October 2018, Catherine, working at the US embassy in Guangzhou, woke in the middle of the night with extreme pain in her temples. She could hear an unfamiliar low oscillating sound. She was aware of the Cuba attacks, although not of the details. Her symptoms were similar to, but not the same as, the others. She began to complain of daily headaches and tiredness. She vomited and had nosebleeds. With Catherine, the syndrome evolved as Hacking described – she was a new person and the two-way loop between her and Havana syndrome brought new symptoms to the disorder. Her worried mother, who flew to China to support her, also became sick. They heard a high-pitched sound in one room and a low pulsing sound in another. Sometimes, it caused them to be abruptly paralysed. Catherine developed hives all over her body and became light sensitive. Unable to bear it, her mother left China after three weeks, and Catherine was evacuated shortly after. Both were given a diagnosis of 'traumatic brain injury.' In total, sixteen embassy staff in China developed symptoms, although most of those were never confirmed as typical cases. Perhaps the symptoms had changed beyond that which the classification could absorb.

I see many patients who struggle to accept the concept of psychosomatic or functional illness and who cannot shake the idea that there is another explanation, whether that's a virus, a toxin or a yet-to-be-discovered disease. In the wake of no better answer or treatment, I ask them to do just one thing – to consider the functional diagnosis and give it a fair chance. It's a treatable condition, and the only harm in considering it is that it could attract the judgement of others. For many people, that feels like too much of a risk, so they reject the diagnosis and go back to looking for their own personal sonic weapon. Some go on to spend a lifetime in that pursuit. In an interview, Catherine's mother reported that she believed both her and her daughter's 'brain injuries' were permanent. In other words, two women are potentially disabled for the rest of their lives by a weapon and an attack that nobody has been able to prove exists. When asked about the possibility of psychosomatic symptoms by her interviewer, Catherine's mother replied, 'there is no way you can fake this.'

6

A Question of Trust

Prejudice: *Unreasonable opinions, without knowledge, thought or reason.*

As the taxi pulled to a stop at the top of the hill, I could see a group of people seated in a semicircle. They sat under a makeshift corrugated awning that was balanced on poles stuck into the dusty ground. Most of the people were dressed in careworn T-shirts and shorts; some wore panama hats, others sported cowboy boots. They sat in couples, with the occasional person sitting alone, and there was one group of three. Chickens pecked the ground around their feet and a saddled donkey was tied to a nearby fence. No one looked happy, and they barely turned their heads to acknowledge the car that was pulling up beside them.

This was La Cansona, a region of Colombia that lies in the heart of the Maria Mountains. It is roughly three hours' drive from the picturesque tourist town of Cartagena. No tourists visit La Cansona, but not because it isn't beautiful – it is. Behind the people gathered on the hill was a ridge, beyond which I could see miles of verdant hilltops and valleys. In another country, one without a tortured history, this would be a place for weekend retreats and country walks, but La Cansona has a violent past that it is still trying to shake off.

I was in the region to talk to schoolgirls who were caught up in a health crisis that had begun in 2014 and was still ongoing. The girls had been told they had mass hysteria, and they were just as angry at the label as any US diplomat would be, but they didn't have a strong political machine to act as a voice for them. In the front passenger seat of the taxi was Carlos, a stocky middle-aged farmer and father of one of the sick girls. He had agreed to be my chaperone for the duration of my stay in El Carmen de Bolívar, a bustling but faded colonial-style town, and the biggest town in the region. Also with us was Catalina, an interpreter. Like my Kazakh companion, Dinara, Catalina is a journalist. Colombian-born and multilingual, she had organized the logistics of my visit, including tracking down Erika Garcia, a resident of El Carmen, who had arranged introductions with some of the families caught up in the outbreak. As an activist for many social and political causes, Erika had set up a support group for the girls. She and Carlos campaigned together. He had arranged this hilltop meeting, but had not told us much about it. Catalina appeared to be as surprised as I was at the sight of the surly looking group.

I had been in Colombia for just a few days, but, by the time I arrived in La Cansona, I had already spoken to several families affected by the outbreak, all of whom lived in El Carmen de Bolívar. While there, Erika had asked me to visit the remoter villages, whose inhabitants couldn't afford to travel to El Carmen. I was happy to make the journey, but had expected to meet the girls one by one, in their homes, as I had done in the city. The semicircle of people waiting for me looked more like an inquisition.

Carlos got out of the car and started greeting the people, leaving Catalina and I behind uncertain about what was expected of us. The car shivered as traffic whipped by at speed.

'What's happening?' I asked Catalina.

'Let's find out,' she said.

We tentatively got out of the car and approached the group, but nobody stood or greeted us. Other people watched us curiously from nearby rustic houses, with only the children and animals brave enough to come in for a closer look. Even when we were within greeting distance of the group we had come to talk to, Carlos didn't offer any introduction. With no obvious social cues to guide me, and no sense of who the leader might be, I tried a smile and started to move from person to person, introducing myself. One by one, I shook the hands of the puzzled people and said my name. Most of the group were in their thirties or forties. There were two younger girls in their late teens or early twenties and I guessed they must have been victims of the illness. The others were almost certainly parents. When I had finished my round of awkward handshakes, Catalina and I sat in the two white plastic seats that stood waiting, facing the rest of the group.

So far on my trip, I had spoken to several affected girls, taking medical histories and hearing their accounts of fainting and convulsions first-hand. The La Cansona group was different because I wasn't able to speak directly to the girls themselves. Faced with a group of couples, all of whom were completely silent, it was hard to know where to start. An angry-looking man in a red shirt took the pressure off me by speaking directly to Catalina, who translated for him. The sound of motorbikes, chickens and dogs was so deafening that I had to lean in to hear.

'We expected you two hours ago,' the man said. 'We've been waiting. People have already left. We've been living with this for nearly six years. We weren't sure we even wanted to meet you. Nobody can be trusted. Maybe we don't want to talk to you now.'

He spoke with an air of authority. The others remained silent, seeming happy to have him as their representative. He was one of the older people in the group. He wore frayed suit trousers that had seen better days, but must have looked very smart when they were new. By his side was a young woman, who I guessed was his daughter.

Unbeknown to Catalina and me, Carlos had promised the villagers that we would meet them at an appointed time – not something he had shared with us. We were staying in a basic hotel in central El Carmen de Bolívar. It had no restaurant, so Carlos took us to a nearby supermarket coffee shop for most of our meals. We had come to enjoy a morning ritual of picking weird treats from a display case full of deep-fried goods. That morning, as we drank coffee and ate buñuelos, or cheese doughnuts, in our cheap supermarket seats, I had noticed Carlos pacing. Only as I sat on this mountain ridge did I discover the source of his discomfort. A kind and deferential man, it seemed he had been too embarrassed to tell us we would have to skip breakfast that day because we had an early meeting.

Carlos's daughter was more severely affected than most of the girls. I had already learned how deeply offended he was by any suggestion that his daughter's problem could be related to hysteria, but still he tolerated me when I asked the families about psychological matters. His manners wouldn't allow him to disagree with me in public, or to rush me over breakfast, so I hadn't known this group was waiting for us under a corrugated roof in the heat. No wonder they seemed so upset.

The man in the red shirt talked very quickly. Catalina couldn't possibly translate every word he was saying, so she attempted to summarize for me. 'He's deciding if they'll talk to you,' she whispered.

Looking at the reluctant, stony-faced group, I decided I didn't want to force people to talk. What I really wanted to do was get right back in the car and drive away.

I whispered to Catalina, 'Please don't try to force them to speak to me if they don't want to. Tell them I'll go for a five-minute walk, so they can talk among themselves about what they want to do.' As an afterthought, I added, 'Although, since my Spanish is so weak, they could be as disparaging as they want about me while I'm sitting right here. I won't understand a word.'

I don't know exactly what Catalina told them, but a few of them started to laugh and I could feel the ice crack. The man in the red shirt spoke again.

'They'll talk to you,' Catalina told me. 'They're just upset because nobody tells them anything and they get no help and their children are sick.'

I nodded to the man and thanked him. He directed his next comments to me, rather than Catalina: 'The parents, we need a psychiatrist. We are under terrible stress. The government won't help us. We can't take it for much longer. We are living a nightmare.'

The story of the girls of El Carmen began in a high school, in 2014. A group of girls, all in the same class, collapsed. Some just flopped to the ground as in a faint, some had convulsions. The condition spread like wildfire. Within a day, girls in several other classes were involved. Emergency calls were made to parents, asking them to collect their children from school, and groups of convulsing, unconscious girls had to be driven in trucks and cars and on the back of motorbikes to the hospital. The phenomenon was dramatic enough to make the news. The girls were pictured in their school uniforms, lying on the ground, inside and outside the hospital. Ambulances and cars were filmed pushing through the crowds, bringing new victims to an already chaotic scene.

I heard of the incident for the first time while speaking at a festival in Colombia, in 2016. Audience members asked me what I thought was wrong with the girls in El Carmen. Convulsing schoolgirls are a key part of the quintessential story that people associate with the acute form of mass hysteria. Typically, one or two girls faint, possibly in the heat, and the rest are triggered to collapse through fright or hyperventilation, or pure expectation. Usually, the phenomenon is gone in a day. I was surprised, then, to be told that the outbreak in El Carmen was ongoing, two years after it had started. I had no answer for the audience members who asked me about it. It was odd, I agreed.

Three years later, five years after the first girl collapsed, I learned that the outbreak still hadn't stopped. On the contrary, it was evolving: new people, new symptoms. By 2019, it was estimated that, out of a population of 120,000 people, as many as 1,000 girls were sick. A phenomenon that had begun in one school had spread to others. A huge medical and social crisis had hit El Carmen de Bolívar and it showed no sign of coming to an end. Most fainting schoolchildren get better, so why hadn't they?

Frida was the first of the girls that I met. All I knew of her ahead of that meeting was that she was one of the lucky ones. She had recovered enough to go to college, while many had not. She was nineteen years old and had moved away from El Carmen to Baranquilla, a coastal town, where she had started college a year late because of how ill she had been. She had kindly agreed to come back to El Carmen to meet me, along with her mother, Jenny. Before talking to her, I had no sense of what the El Carmen outbreak was about, but she began to clear up the mystery very quickly.

We arranged to meet in a restaurant Carlos had chosen,

in central El Carmen. It was quiet, and furnished with plastic chairs and tables covered with frayed salmon-pink tablecloths. Catalina and I ate dinner together while we waited for Frida and her mother to arrive. Carlos never ate with us. For the duration of my stay in the town, unless I was in my hotel room, Carlos was never more than a few feet away, although he was often so quiet I could almost forget he was there. As we waited for Frida, he sipped water and watched me. We were the only customers in the restaurant. Every time the door opened, I looked to see if it was Frida, but most of the time it was just the waiter. When Jenny finally appeared alone, later than I was expecting, I was worried that Frida had changed her mind.

'She's here,' Jenny reassured me.

Frida was outside, cleaning her trainers, which were dirty from the walk to the restaurant; she didn't want to meet us until she had removed every trace of dirt. Several minutes passed before she appeared, her trainers gleaming white. She reminded me of my niece, who has exactly the same priorities when it comes to trainers – they cannot be white enough. Knowing Frida had been very sick, it came as a great relief to find a healthy-looking young woman with the same concerns as any other teenage girl. Tall and willowy, Frida's skin was glowing and her thick black hair shone.

There were quick introductions. To my surprise, before Frida was even settled in her seat, Catalina pushed my plate of leftover dinner in her direction. I asked Jenny if I should order something for them, but she said no. Meanwhile, Frida took my plate and began wolfing the food down with a laugh. Nothing about the way she ate or held herself suggested she had recently been very sick. I was struck by the immediate rapport between Frida and Catalina, and realized I had a great deal to learn about Colombian culture.

With Frida distracted for a second time, having moved from dirty trainers to food, I fell into conversation with her mother. At first, we only indulged in small talk – about the rain, the traffic, the town. Frida didn't appear to be listening. Once, when I lumbered into a question that was too personal too soon, Jenny put me right: 'Not yet.' It was only when Frida had cleaned her plate and her mother had finished testing me that I was told it was okay for us to start talking about the events of 2014. As soon as the subject opened up, Jenny made it very clear how hard their lives had been for the previous five years. She started to cry and told me that she had two daughters who were affected.

'I thought I would see my daughters die on the ground,' she said. 'The newspapers and the doctors, they took away my daughters' dignity. They called them crazy.'

In 2015, the Instituto Nacional de Salud (National Health Institute) of Colombia (the INS) published the results of a detailed epidemiological study, declaring that the outbreak was due to mass psychogenic illness (MPI), the modern term for mass hysteria. This label had hit the families hard – every parent would tell me this.

'I'm sorry,' I said. I looked at Frida, but she did not seem in any way upset. In fact, she was looking me up and down with interest while her mother talked. 'Is it okay if Frida tells me her story in her own words?' I asked Jenny.

It would become a feature of the interviews that I had to ask the parents to let the girls tell the story. When I did, the mothers and fathers happily pushed their child to the fore, but it was never their first impulse. At my request, Jenny deferred to Frida, who pushed her plate away and stared at me even more intently.

'Where do you come from?' she said, changing the course of the conversation.

I told her and, in return, I asked her to write down her full name for me. Frida, it turned out, was not even her name – that was Frinia. She made people call her Frida because she preferred it.

'After Frida Kahlo,' she said, before adding, 'because I love her.'

Frida Kahlo was her hero. The name change turned out to be apt, because painting had been key to her recovery.

'Do you know where the outbreak started? In which school?' I asked. I wondered if she knew patient zero.

'It started at my school,' Frida said, 'in my sister's classroom, next door to mine.'

In subsequent conversations, other girls would identify other schools as the starting point. Both the narratives around outbreaks like these and the symptoms themselves tend to evolve as they spread. It's essential to know the stories that are circulating, even those that are no more than rumour, because, like the threat of a sonic weapon, they are the driving force of the phenomenon. Memory is tricky; stories evolve in the telling and retelling, with contributions from many parties. I had to accept that not everything the families told me would be accurate. But they were truthful to their own experience and this was what I was after – an accurate portrayal of what it felt like to be on the inside. The girls' and their families' personal experience was what propelled things forward, even if what they told me was not strictly true.

Frida recalled that it had been a very hot time of year and the girls had been complaining of feeling vaguely unwell for a while. It was a mixed-sex school, but there were more girls than boys overall. The windows in her sister's classroom didn't open properly, so it was unbearably hot in there at times, with fifty pupils in a confined, overheated space.

One sticky, uncomfortable afternoon, Frida's class heard unusual noises coming from next door. They looked out into the corridor and were surprised to see girls being carried out of the classroom. She learned later that one girl had complained of being unable to breathe, becoming increasingly distressed until she seemed to faint. Almost immediately, fifteen other girls collapsed. The pandemonium that ensued was enough to alert the whole school. Classes stopped and people poured into the corridors. Parents and ambulances were called and girls were rushed to the hospital en masse. These seizures then quickly spread to other classes – Frida estimated that, by the end of the first day, at least fifty girls had collapsed.

'Did you see people collapse?' I asked her.

She described girls trembling and shivering, lying on the ground with their eyes closed. She remembered one girl with her face pulled to one side and her mouth twisted.

'Sometimes they stayed sitting down and kept shaking, but could still talk,' Frida told me.

She showed me a video that I had seen before, recorded by a news station. Young women had been filmed in various stages of collapse: some lying still and pale, others with backs arched; some writhing, some shaking and several being restrained by male relatives or boys from their class. These were clear dissociative seizures. There was no question.

'You didn't collapse?' I asked.

'There are two types,' Frida said. 'Those with seizures and those without.'

The first day, all the girls had seizures, and Frida referred to that as 'the catastrophic day'. However, by the end of the week, more than a hundred girls were sick, and not only with seizures. Over time, the illness evolved into a whole range of symptoms, from visual blurring and dizziness to breathing difficulties, headache and chest pain, while some girls had more insidious problems, like poor skin and hair loss. As new cases emerged, some still had convulsions, but not all did.

Frida did not have seizures. She described her symptoms in detail to me, many of which had been transient and non-specific. She had developed digestive problems, knee pain, visual disturbance, weight loss and skin problems, each of which came and went in a creeping fashion. At her worst, she struggled to go to school, but most of the time she managed to carry on. Although she did not have a specific diagnosis, she had been given a cocktail of medications.

'I didn't always take them,' she told me.

I asked about the healthcare structure in Colombia and was told it operated under an insurance system. Those with exceptionally low or no income didn't have to pay for insurance, but most did. There were different insurance companies offering different levels of care. Insurance companies ran their own hospitals and many doctors worked for specific insurers. By Jenny's account, most of the girls that she knew of had been investigated fairly thoroughly. They'd had blood tests, brain scans and EEGs. After the early tests had all come back clear, they were ultimately screened for heavy-metals poisoning and rare diseases. Those tests were also normal.

As Frida told me about the girls' medical assessments, she did not seem disappointed with the thoroughness of the testing, but she clearly thought there was a problem with the results.

'They did tests, but they didn't tell us what they showed. They did different tests on different girls, as if they were acting randomly, with no plan. Sometimes, blood tests came back and the results were identical for every girl. How could that be?'

This issue tempted Carlos out of his deferential silence and he interjected, 'The government spent 700 million pesos on research, but then didn't give us all the results.'

He was referring to the INS study, suggesting the final conclusion did not include all the data. Frida, Carlos and Jenny all agreed that information was being withheld and that their concerns were not being taken seriously. There was a definite air of suspicion.

'The Red Cross came, but the government sent them away,' Jenny told me.

That seemed unlikely to me, but the rumour spoke to the lack of trust between the families, the insurance companies and other authority figures. It was almost the exact opposite of Havana, where medical professionals and politicians held such sway.

'In the end,' Jenny said, 'we didn't know what was true and what was fiction. They said some children were sick because the fathers worked with heavy metals. But why wasn't the father sick, if that was the truth?'

'Quite right,' I agreed.

Many of the causes that had been suggested to Jenny and Frida made no sense. Each suggestion only served to frustrate the families further. But it was when the president of Colombia made a public announcement stating that the outbreak was due to mass psychogenic illness that the community became really upset.

'People in the street shouted after our daughters – "There go the crazies!"'

The families took the label hard and were convinced their children's medical care declined as a result of the diagnosis. Girls presenting to the emergency department with seizures were put in a special holding room instead of getting immediate attention. As a doctor who treats dissociative seizures, I was conflicted when they told me this, since conservative care that avoids excessive tests and drugs is the right approach for these sorts of attacks. But that care must be explained properly to those involved so that it doesn't look neglectful. Clearly, the parents did not see conservative care as good care, but I could not tease out where the breakdown in relationships began.

'If I went to the doctor, all they did was offer me acetaminophen [paracetamol],' Frida told me.

'They love acetaminophen in this country,' Catalina interjected, between translations. 'You get it for anything.'

'Did they give you any diagnosis that made sense?' I asked Frida.

'No.' She shook her head.

'Are you well now?' I asked her.

'I feel well, but I know I'm not well. I know it's still inside me,' she said.

The comment chilled me. Over the coming days, it would sum up exactly how the girls had come to feel about their health – and, to my mind, explained why they were not getting better. It was as if she had become accustomed to the expectation of ill health.

'If you feel well now, maybe it's gone,' I countered, hopefully.

'In the morning, I have to get a bus to college. I never get a seat. The bus is already too busy when it gets to my stop. I have to stand, and it's a very long way. By the end, I always feel dizzy and sometimes sick.'

She had been so scared, she seemed to have lost the ability to trust her own body. The feelings of faintness induced by a long, overheated bus journey sounded like a normal thing to me, but, for her, they had a greater significance. I suspected that her focus on the 'white noise' was making it hard for her to believe in her own recovery.

Frida didn't progress to seizures, but her sister did – she had fainted during the 'catastrophic day'. Frida and Jenny described that first day's hospital experience: the emergency room wasn't big enough to contain so many casualties arriving at one time. There wasn't even enough space on the floor to accommodate everyone, so some girls lay in their parents' arms, still convulsing.

The school closed for three days after the incident. When it reopened, some parents were reluctant to send their children back. The acute phase of the problem was over, but people were frightened, and the sickness didn't stop. New victims continued to emerge, the original girls failed to recover and other schools in the region became involved.

It sounded very much to me as if the original class had suffered from being in a poorly ventilated classroom. A girl fainting in such an environment would not be so very odd, and she might well provoke the others into doing the same thing. But what was sustaining an event that should have run its course years before?

'Why do you think it happened?' I asked Frida.

'It was the HPV vaccine,' she told me confidently.

And there it was: the recurring theme that would colour, and sometimes completely take over, every subsequent conversation I had in the town; the illness vector that was stimulating the outbreak, keeping it alive. The people of El Carmen had come to believe that the HPV vaccination had made the girls ill. When Frida told me she knew the illness was still inside her, she was referring to the vaccine. She believed she had been poisoned. It is astonishing how often illness is attributed to a malign force – poison in Krasnogorsk, a sonic weapon in Cuba. Conspiracy is more compelling than normal life. External causes for illness are more attractive than psychological mechanisms. It is also soothing to have a single tangible explanation for illness, rather than a chaos of unknowns.

'When were the girls vaccinated?' I asked.

Children fainting due to a vaccination or blood test is common. Painful stimuli or sudden fright activates the autonomic nervous system in such a way as to cause an abrupt fall in blood pressure, and collapse. This is called reflex anoxic syncope, and it is an immediate physiological reaction, which doesn't indicate any underlying disease and has nothing to do with psychological trauma.

'I had the second dose a month before,' Frida told me.

Not reflex anoxic syncope, then.

'And the girls in your sister's class?'

'Also a month before.'

'Okay.'

I asked Jenny and Carlos if they also associated the outbreak with the HPV vaccine, and they assured me they knew for a fact that the vaccine had made their children sick.

'But how could a vaccine given weeks before cause so many faints, on a single day, many weeks later?' I asked, unable to contain my surprise. 'Surely it was caused by something that happened that exact day?'

Recall bias causes us to connect events that happen in close proximity, but proximity does not prove cause and effect. To say that a sound caused brain injury, the biology has to make sense. Similarly, one has to know how a vaccine could create the clinical syndrome of mass seizures.

'We've been told this is happening all over the world. In Japan. In Italy,' Carlos told me.

Where had this information come from, and how was the connection first made between the catastrophic day and the HPV vaccine? Why did Frida blame the vaccine?

'There were problems with the vaccine batch that I was given,' she told me. 'The man who brought the samples to the town was drunk. He brought them on a motorbike, but didn't stay with them. He didn't keep the vaccines cold.'

She went on to say that the vaccine had been administered by a nurse who wasn't wearing gloves or a mask. The nurse had left the door of the fridge open between patients. The girls' arms were bruised and some were swollen the next day.

But all of that was weeks before the first seizure. I couldn't escape the fact that the girls collapsed one after another, minutes apart. Frida's answer didn't acknowledge that.

'It sounds like the vaccination programme wasn't well managed,' I said.

'They didn't even ask the parents' permission,' Jenny said.

'That doesn't sound right, and I can see why it worried you. But why do you think fifty girls had seizures simultaneously?' They all looked puzzled, but it seemed key to me, so I pressed on: 'If a group of people were all exposed to a flu virus on the same day, they would expect to get sick at different times, to different degrees. But these girls got sick at the same moment. Why do you think that was?'

Frida was the one who finally answered. 'It was very frightening, in the classroom. If you saw somebody collapse, you felt you would too. It spread like that.'

I looked at Jenny and Carlos to see how they would react to this. They didn't. I repeated what Frida had said: 'So, people collapsed because they were so frightened, because they had seen other people collapse, right?'

Frida nodded in agreement, adding, 'It was awful seeing people collapse.'

I turned to Jenny and Carlos again. I hoped they would see the wisdom in what Frida had just said. It didn't work that way, though. Instead, Carlos redirected the conversation, telling me that a Dr Herrera, an anthropologist, had confirmed the outbreak was due to the HPV vaccine, and that it was killing girls all over the world. Carlos spoke very little during my visit, but he became animated when discussing a small number of topics – among them, the HPV vaccination and Dr Herrera.

'The HPV vaccine isn't killing girls. That simply isn't true,' I said, shaking my head, but I could see that neither Jenny nor Carlos were very impressed to hear me say that. I asked where they had found this information, and learned that Dr Herrera was at the heart of it. A Colombian national who lived in the USA, he had come to the town when he learned about the outbreak. I don't know if he personally introduced the idea that the vaccination programme poisoned the girls, but he certainly encouraged it. When I asked who this influential anthropologist worked for, none of them knew.

'I'm not sure Carlos understands that Dr Herrera is not a medical doctor,' Catalina whispered to me.

'They seem to put a lot of trust in him.'

'Carlos keeps saying that he's the town's saviour.'

There was no pathological mechanism through which a vaccine could have caused a mass outbreak of seizures, on a single day, a month after the vaccine was given. There is no association between the HPV vaccine and epilepsy, and, even if there was, the seizures I had seen were dissociative and certainly not caused by a brain disease. But I didn't know these people, or Dr Herrera. I decided not to challenge them further on this topic, and to continue to try to understand the girls' experience.

'What made you better, Frida?' I asked. Even if she hadn't fully recovered, she was moving on.

The turning point came for her with the intervention of another visitor, Mila, who Frida told me worked for Médecins

Sans Frontières. 'She held workshops for us,' she said. 'We painted and she taught us how to control our breathing. She talked to us about diet and gave us relaxation exercises.'

It seemed that Mila's intervention, and the painting in particular, had been very beneficial for Frida and for the other girls who could access it. Mila ran all-day workshops, provided the girls with lunch, and her organization had paid for a bus to collect the girls and bring them for treatment. This mattered a great deal, to the poorer families especially.

'There were six girls in a workshop,' Frida told me. 'You couldn't have more than six, or girls would start to have seizures.'

For reasons unspecified to me, Mila had left after three months. She had only been able to see a small number of girls. Frida told me she was due to return, and that she thought Mila was writing a thesis about the phenomenon.

'I thought you said she worked for Médecins Sans Frontières?'

'I'm not sure.'

'Do you know her last name?'

They didn't. Like Dr Herrera, Mila was a shadowy figure, albeit one who sounded like she had really helped the girls.

'Did the girls get much psychological support from elsewhere? From the insurers?' I asked.

The government had sent psychologists, it turned out. Fifty girls were brought to see them, all on the same day. Thirty of those girls collapsed simultaneously at that meeting. It had been yet more pandemonium. The psychologists were supposed to stay for five days, but left after three. Nobody wanted them to come back.

It seemed to me that a lot of people were horribly out of their depth in El Carmen. Not just the girls and their families, but the medical teams too. Good medical care is available in Colombia, but no emergency room in any small hospital, in any country, is prepared for fifty convulsing girls to arrive through the door in a short space of time. I don't think a UK hospital would have been any less overwhelmed. But it wasn't the acute care or the disappearance of the psychologists that upset the families the most; it was the impact of the diagnosis of mass psychogenic illness. Once that had been published, numerous hurtful accusations and hypotheses were levelled at the girls.

'People said they needed husbands,' Jenny told me.

How very nineteenth century. A hundred years ago, people blamed these sorts of seizures on the ovaries, on the womb, on too much sex, or too little. Treatment prescribed ranged from masturbation to hysterectomy, ovarian massage and less sex – or more. Every time I dare to hope these sorts of attitudes have been left behind, I find evidence to the contrary. They are not confined to Colombia or South America. They linger, in varying degrees, throughout the world.

Another theory was that the children were inbred, since so many of them had the same last name.

'How can that be?' Jenny said, angrily. 'Why would only the teenage girls be affected, if we are all inbred?'

She was right; it was a nonsense. It was becoming clear to me why the diagnosis of mass psychogenic illness had been rejected so completely. The oldest and most pejorative interpretation and associations that go along with this label had been directed at the girls repeatedly. If any attempt had been made to explain it in a contemporary context, they hadn't heard it, and they were now resolutely closed to the suggestion. They batted it away, along with all the other insulting explanations they had heard: they were sick because they used Ouija boards; because of poor nutrition; because they ate too many fries.

'Eating too many fries!' Jenny scoffed.

The girls had been demeaned by public conversation. Feeling isolated and duped by authority figures, the families applied their common sense to dismiss most of the theories. The only explanation that seemed to link the girls was that they had all received the HPV vaccination. And, of course, that was a more attractive explanation than MPI, with all its connotations. The dubious batch of vaccine that a drunk delivery man was said to have left lying around in the heat only added to the appeal of this theory.

'But not all the girls were vaccinated by that batch,' I pointed out, at which point Dr Herrera entered the story again.

'He came as a godsend,' Carlos said, putting his hands together in prayer.

Dr Herrera had the answer. He had taken the girls' story to a medical symposium in the USA and reported back to the families that the doctors at the symposium had cried when they saw the videos of the seizures. They had unanimously agreed that they were definitely caused by the vaccine.

'What sort of doctors were they?' I asked, bluntly – perhaps excessively so. I couldn't hide my surprise at this strange account of weeping doctors. Doctors cry, but not at academic conferences. Carlos didn't know the details, but advised me I could watch Dr Herrera's lecture on YouTube. I had no access to the Internet in El Carmen, so would have to wait to do more research.

It was clear that nobody really knew the details of Dr Herrera's qualifications or affiliations – or Mila's, for that matter. I asked them what they knew about my qualifications. When I had contacted the group, I had introduced myself, including links to my NHS work and my publisher's website, so they could verify my identity and credentials. Erika and I had discussed (via Catalina) the purpose and limitations of my visit, at great length, before it was agreed. Even so, when directly questioned, Jenny, Frida and Carlos knew as little about me as they did about Dr Herrera and Mila. Why were they so open with me? Probably because I am a doctor and a foreigner to them, and, of course, because they were desperate for help and for people to know of their plight.

We talked for a couple of hours before the evening drew to a close and we walked together out into the street. As we said goodbye, Frida and I hugged. I tried to encourage her to enjoy being well.

'It was so hard to make the decision to send her and her sister away to college,' her mother told me as we stood outside the restaurant.

She had delayed a year before sending the girls to Barranquilla, and then took a leap of faith.

'But you did the right thing,' I said. 'Look at her – she's thriving. She's moving forward.' I wanted them to worry less. Frida stood beside us, beaming, and I felt hopeful.

Back in my hotel room, I sat thinking about everyone I had spoken to that day. Frida had been intelligent and funny. In between my questions to her, she had peppered me with questions about myself and was full of interest in the answers. Jenny was a worried, frightened parent, who seemed to be suffering more than her daughters as she worked hard to preserve their dignity. Carlos was harder to read. Most of the time, he just let the conversation happen, hiding his thoughts behind a half-smile. I knew his daughter was much sicker than Frida, but he was clearly less able to talk about it.

I also thought a great deal about those people I had not met: Dr Herrera and Mila. Who were they? It felt odd not to have access to an online world in which to search for instantaneous clues. Perhaps sending the families webpage links confirming my identity had been a waste of time, if only a small proportion had access to the Internet and few could speak English. I was learning that relative isolation from the wider world forced the people of El Carmen to choose who to trust – and who to mistrust – at face value. Colombia's history had taught them to be suspicious of home-grown authority figures, and the country's turbulent past meant there were few outside visitors. Visiting professionals had not had the chance to tarnish their reputation, and seemed inherently more trustworthy, I guessed. My guilt at my own visit began to eat away at me.

The next day, as I was putting on my trainers, I found myself laughing. I was thinking of Frida. Once we had relaxed into each other's company, I hadn't been able to resist teasing her about how long it had taken her to clean her trainers to her high standards. Responding in the same spirit, she advised me that I would soon be eating my words, promising that, where I was going next, I could expect rivers of mud. She mimed me wading gracelessly through water up to mid-calf, and assured me I would be equally dedicated to trainer-cleaning in the days to come. I was relieved, therefore, when we arrived at our next meeting place and I discovered that she had been exaggerating the magnitude of the mudslide.

We had arranged to meet Juliet, the next girl on our list of those affected by the outbreak, at her home, on the outskirts of El Carmen, but, to get to her house, we had to abandon the car on the main road and walk uphill. Road or sewer works had caused temporary drainage problems and the streets were thick with mud, slippery and potholed – although not a river, as Frida had promised. Even so, we had to hop from dry patch to dry patch, with the occasional squelching misstep. Catalina and I got to our destination relatively clean, with Carlos walking anxiously behind us.

Juliet and her mother, Yeliza, were waiting for us outside their home. It was a low makeshift building, largely a frame, without a door or windows, and, as I discovered when I stepped inside, no flooring – there was only dry, bare ground, swept free of dust and loose dirt. Gaps under the walls suggested that, when it rained, water must come through the inside of the house, and there were pots on an open fire out the back. It was much more basic than any of the homes in central El Carmen. Later, when I asked Catalina about the family's living conditions, she assured me that Juliet's family had electricity and were almost certain to have modern conveniences like a washing machine. There was stereo equipment and a fridge in the room where we met, but I couldn't see where they plugged them in. There was also a large blue barrel filled with water, and, beside it, a bucket filled with cassava. Three chairs sat awkwardly out of place on the dirt floor, waiting for us.

Juliet was a tiny eighteen-year-old who looked much younger, and she was heavily pregnant. Like Frida, she was dark-skinned and beautiful. As in most South American countries, people in Colombia are often mixed ethnicity; influxes of slaves and traders and conquerors had brought with them a variety of cultural influences and given their descendants a wide range of appearances. Catalina was from Bogotá and she was as pale as me. She had studied in England, where people didn't believe she was Colombian because she didn't look like their idea of a South American person. All the girls I met in El Carmen had the sort of beauty that is coveted in European and North American cultures, with dark hair and eyes, and fine features. They were all fresh-faced, wearing no make-up, and were careful in their appearance.

Once we had sat down, Juliet began to tell me her story. Yeliza stood behind her daughter's chair, a hand touching her shoulder. Again, I had asked for Juliet's personal account. I was trying very hard to see through the people's interpretations to the experience itself, as I would in a medical consultation. I kept reminding myself not to assume a diagnosis until I had heard each story.

Juliet's seizures had started in school. Not in the same school or the same wave of attacks that Frida had described, but months later, on Ecology Day. She hadn't been feeling very well. It was hot and she had been standing for a long time at a ceremony. She was dizzy and light-headed, and began to notice spots in front of her eyes. Suddenly, everything went black and she flopped to the floor. She woke quickly and didn't convulse. Her description was entirely consistent with a faint.

When Juliet collapsed, there was no panic. Her friends coped well. They brought her home and she had appeared to recover. Three days later, she complained of heaviness in her chest and collapsed again. After that, the collapses became a regular occurrence, but the manner of them changed.

In church one morning, she lost consciousness. 'I felt very tired. I knew I would collapse. I know, if I feel that way, that I can sometimes stop it by calming myself down. That time, I tried closing my eyes, but it didn't work. Everything went black and I woke on the floor.'

People told her that she had convulsed, then she lay still for a while, as if asleep. When she woke up, she was crying. It is very common for people to cry in association with dissociative seizures. Like Lyubov, Juliet did not find it easy to connect this ordinary physical indicator of upset with its most common cause.

'Do the seizures have a pattern?' I asked.

She thought for a while before saying, 'They happen when I think too much.'

So many of the girls associated their symptoms with anxious thoughts, but they then overlooked the importance of that and searched for alternative explanations. Despite being aware of a strong link between the content of her thoughts and her symptoms, Juliet still rooted the problem in poison, insisting, as Frida had, that the vaccine was the cause. The vaccine narrative had taken a very strong hold of the town and eclipsed every other possibility.

Both Juliet's parents worked, but still they were not well off and so had access to free healthcare through a government insurance company. Juliet had a range of tests when she got sick, including brain scans, EEG, blood tests and cardiac tests. They all came back normal, and eventually she was referred to a psychologist. By Juliet's recollection, most of the psychologist's questions were focused on her upbringing. Juliet was left feeling as if her family were being blamed. She also got a sense that people believed she was making up her symptoms.

As Juliet told me how she felt, her mother started to cry, saying, 'As a mother, I'm the one who should have had a shrink.'

Almost every parent I spoke to echoed this thought. It was the most curious thing. All the parents felt the need for psychological support, but they refused to countenance any suggestion that their daughters might need the same. It was as if they could not accept that a child has their own emotional world, at least as fragile as any adult's.

Before I could ask Yeliza more, we were interrupted by a shy-looking girl who appeared in the doorway. A chicken made its way into the room behind her. Yeliza introduced the girl as her younger daughter, Paula, who was also sick.

'You have seizures too?' I asked her.

Her mother answered for her: 'She has very painful periods. Very heavy.'

I enquired about the symptoms Frida and Juliet had told me about: dizziness, visual disturbance, fatigue, joint pains, fainting. Paula had none of these, just painful, heavy periods.

'It sounds a little bit different from the other girls,' I ventured.

'I know what periods should be like, and hers are not normal. I'm worried,' her mother told me.

I wondered aloud if perhaps Paula had painful periods for another reason – either as part of a normal spectrum, or as a consequence of a gynaecological problem.

'But she had the vaccine too,' her mother countered.

Then came the story I had heard before, about the man who had delivered the vaccine drunk, who had broken the cold cycle. She told me how little trust she had in the insurance companies, in the government and in the doctors they had seen. She believed the local doctors knew that the vaccine had been faulty, but were not allowed to admit it. When I asked why she was so certain of this, she repeated what Carlos had told me about HPV vaccine killing girls. I made an attempt to tell her this wasn't true, but she started to cry. I felt cruel.

'One doctor tried to claim my daughters never had the vaccine, but they did,' Yeliza said.

She didn't know who to trust and didn't feel well served by her insurers. A promise that they would refer Juliet to another doctor had never materialized, while Paula hadn't seen a doctor at all. They felt abandoned. With all the talk about a poisonous vaccine and a lack of trust in official reports, it's hardly surprising that Yeliza was terrified on her daughters' behalf.

I was sure Juliet had fainted and had subsequent dissociative seizures, or even panic attacks. From the description, all of the appropriate tests had been done and she had been given the correct diagnosis. But that's when things had fallen apart. The problem was their understanding of the diagnosis.

This could have been a result of how the news was delivered, or it could have been entirely down to the reputation of psychosomatic conditions. The same thing happens everywhere, with every doctor who deals with dissociative seizures – it is a difficult diagnosis to communicate, and it carries with it centuries of misinformation. By Juliet's account, the discussion with the psychologist had placed all the emphasis on poverty and her upbringing. But that made no sense to her; nothing had changed in her life, and her home was a happy one.

'They seem poor to you because of where you're from,' Catalina would say to me later. 'But this is normal life for them. These houses without floors – everybody they know lives in this kind of house.'

The emphasis on poverty and potential abuse had irked everybody. The view that dissociative seizures are always due to abuse, psychological trauma or psychological suffering is limited and old-fashioned. Despite the fact that the illness was sweeping through the town, and not unique to Juliet, the psychologist had focused attention on her family and her personal life, which had alienated her. On the biopsychosocial spectrum, the outbreak in El Carmen was predominantly a social one. It was less about individuals and more about group dynamics. It was about a government that people didn't trust, a flawed healthcare system and people's isolation from the outside world. Rumours were spreading through the town and the families had no easy way to filter truth from lies. That Juliet had found discussions about her family unhelpful was not surprising.

The good news was that Juliet was better. She had managed to find ways to overcome her symptoms, teaching herself techniques to stop the seizures, such as by closing her eyes and telling herself to calm down when she felt the warning signs. Although this worked, Juliet was scared that it wouldn't work forever, echoing Frida's concerns that the illness lay dormant inside her.

We talked for a while about dissociative seizures. In a third of cases, they get better just through having an open discussion about how and why they manifest. Mechanisms like predictive coding and the embodiment of illness templates create disability through expectations, but they can also heal. Juliet had found ways to manage her seizures, and positive expectations could help her to sustain that.

Just before saying goodbye, I decided to push the boundaries of the discussion and suggested to Yeliza that Paula's period pains did not mean she had the same disorder as the other girls. If they were worried, I told them, they should ask their insurer for an appointment with a family doctor or gynaecologist, who would hopefully allay their fears.

As I left, Yeliza began to cry and couldn't seem to stop. 'Thank you. Nobody has ever talked to my girls before,' she said.

Her thanks made me feel guilty, because I knew I was making a passing visit, just like all the other people who had come to the town to write about them for newspapers, never to return. I thanked them for their time and headed down the muddy hill again, knowing I would not be back.

After that, Catalina, Carlos and I moved through El Carmen and its suburbs, collecting stories. Each added something new, while also repeating familiar themes. The girls had been referred to as mad, crazy, attention-seeking, actresses, uneducated, simple and sexually frustrated. They felt traumatized, first by the illness and subsequently by the way they had been portrayed. All the families were under strain, psychologically and financially. None felt they had been given an adequate explanation for what was happening to them.

Marcela had two sick daughters, and we met her with the eldest, Yesmid. The illness had destroyed their lives, they told us. The family lived in Caracoli, a village outside El Carmen. As we chatted, we sat under a trellis dripping with flowers, outside their pretty pink and green home. Yesmid, who was twenty, was nursing her newborn baby. She had dissociative seizures and told us that she couldn't stop crying during them. She didn't know why that was.

What was odd was that Marcela, Yesmid's mother, had started to have seizures too. 'I was overwhelmed. I still am,' she told me, when I asked why she thought it happened to her. She blamed stress for her seizures.

I blundered in, as usual: 'But if your illness was triggered by stress, don't you think that could also apply to your daughters?'

'No,' Marcela said, looking perplexed. 'Why would a child have stress?'

Even her daughter's tears hadn't indicated unhappiness to Marcela. In fact, the whole family had good reason to feel very stressed: they had once been relatively well off, but they'd had to sell their shop and motorcycle taxi to pay for an expensive array of medical tests. Now, they couldn't even afford the bus fare to El Carmen.

'The insurers will pay for my daughters to see a psychologist and physiotherapist,' Marcela told me, 'but they won't pay for the bus to the hospital. So, we can't go.'

She showed me piles of medical notes from the many hospital visits that had financially crippled the family – their insurance had only covered part of their medical care. I found the investigations to be very thorough. Everything that would be done in the London hospital where I work had been done here – although, in London, of course, they would not have had to pay. The diagnosis for mother and daughter was listed as 'non-epileptic attacks', another name for dissociative seizures.

Later, I asked Catalina why she thought the families were so unanimous that stress was destroying the parents' health, but they didn't acknowledge the psychological impact of the circumstances on the children.

'Maybe they think they've protected the children, shielded them more than was ever really possible?'

We agreed they did not seem to accept the maturity and independent inner lives of the girls.

Many of the girls I met during my visit had recovered. Some, like Marjory, who lay languidly in a rocking chair during our conversation, felt they had put the illness behind them, but their parents almost never had.

'I'm afraid to tell my parents if I have any tiny pain, because I know they'll worry,' Marjory told me. 'So, if anything happens, I just keep it to myself.'

Marjory had also benefitted from meeting the mysterious Mila, who had helped Frida.

'Who is Mila?' I asked the family, still trying to clear up the uncertainty left by Frida.

They weren't sure, but they thought she was a psychologist. The parents did not get to know her. She had collected the children by bus, and saw them alone.

'Where is she from?' I asked.

Maybe Bogotá, they thought, although they couldn't be sure.

'What's her last name?' I asked.

Nobody knew.

The person I had hoped to meet was Carlos' daughter, but I never did. She was one of the most tragic victims of the El Carmen outbreak. Her initial symptoms were physical, just like the others, but she had quickly descended into severe psychological distress and had tried to take her own life. She was admitted to a psychiatric institution for fifteen days, and, while she was there, another patient tried to sexually assault her. Men, women and children had been mixed together on the same ward, Carlos told me. He had to fight to take his daughter home. I knew that Carlos was holding onto the conviction that the HPV vaccine was the cause, and hoped that, once this was acknowledged, somebody would offer a cure.

Every now and then, he would gently voice his objection to a psychological explanation for the outbreak. On our way to see Laura, another young victim, he said, 'The next girl has lupus. That cannot be psychological.' This was as forceful as Carlos' opposition to my views ever got. Lupus does not have a psychological cause, but it is easily over-diagnosed, which is what I suspected might have happened in Laura's case. The blood tests used to diagnosis lupus are fraught with error and false positives. Autoantibody tests are difficult to interpret, and I wouldn't usually make the diagnosis without specialist help. Would Laura's diagnosis prove to be sound, or had minor symptoms and borderline blood-test results been over-interpreted?

She was waiting for me, surrounded by her family, in her home in central El Carmen. The house was whitewashed, modern and packed with furniture. The front room was accessed through a door that opened directly onto the street. The door stayed wide open during the day, allowing air to move freely in and out, but it also made the room feel like it was part of the street outside. Walking around El Carmen, I sometimes felt I had direct access to every person's home through their open front doors. It was very different from London, where our homes are where we hide from outsiders.

Laura was another bright, pretty, dark-haired girl, who

appeared to glow with health. She was twenty when we met, studying social communication at university in Cartagena. She hoped to be a journalist. Laura did not get sick with the first wave of girls; for her, it came later, and started with joint pains. A sporty girl, she played a lot of volleyball, and doctors advised her that she had overdone it. She was told to give up sport, which she did reluctantly. Her joint pains got worse. Then, she became short of breath, at times feeling as if she were suffocating. She felt hot all the time and her mother constantly fanned her to cool her down. She developed fevers and hyperventilated regularly. She had pains in her chest and her hair fell out. She developed a rash on her face and, finally, she lost consciousness and had convulsions.

'Look!' Her mother, Kim, showed me a photo of Laura at her worst. Catalina and I gasped. The girl in the picture had no hair, her face was swollen like a balloon. I would not have recognized her as the girl sitting in front of me. In another picture, her face had the florid red butterfly rash across her cheeks that is absolutely characteristic of lupus. Laura handed me a thick pile of medical notes. Everybody watched me expectantly as I read through them: *Severe kidney failure. Dialysis. Multi-organ failure. Pericarditis. Cerebral oedema. Pleural effusion. Severe anaemia.*

Laura had nearly died; there was no doubt that she had lupus. It had taken months for her to get that diagnosis, and it was almost too late. Her family had to pay a private doctor, who admitted her to hospital, where she stayed for four months, in and out of the intensive-care unit. It truly felt like a miracle to see the same girl sitting in front of me, smiling, in the third year of her college course, living a normal independent life.

Kim was a resourceful, tough woman. When the family ran out of money for treatment, she had gone to the governor of the region and appealed for help. The governor organized for Laura to see a new doctor and arranged for all her medical care to be covered by insurance.

'Wow,' I said, genuinely surprised. 'That was kind of them.'

'No,' Kim replied. 'In this country, if somebody gives you special treatment, there is a reason.'

I looked at Catalina for clarification.

'She thinks the governor paid for the treatment because they were hiding something,' she explained.

For a while, we talked about how sick Laura had been, but the conversation kept veering away from her, being pushed in the same direction it always went in El Carmen: towards the presumed cover-up, the corruption and – of course – the HPV vaccine. Kim described an atmosphere of insult and perceived neglect.

'I realize it was slow,' I said, 'but she got good care in the end.' I indicated the notes in my hand.

'Because I could fight for her. Not everybody could do that for their children. They were told that they were crazy because they were born in a place with a violent past. When we asked why it was only the girls who were affected, we were told the violence had entered them through their genes.'

In the past, the area around El Carmen – the Maria Mountains, in particular – had been very dangerous. In 2000, it had been the scene of a notorious incident called the Massacre of El Salado, in which the AUC, the paramilitary group responsible for the most deaths in Colombia in recent decades, launched an attack on the poorest people of the area. Four hundred and fifty men descended on the villages, raping, torturing and terrorizing the inhabitants. The horror lasted over a week, and was so brutal they even targeted children.

The El Carmen girls got sick in 2014. Many had not been

born at the time of the massacre, while some would have only been infants. But they all lived through the years of violence that followed. Violence was a backdrop to their childhoods, but that had also been the case for children all over Colombia, for decades – and for their parents, too. The focus on historical violence made no sense to the families – there was nothing to suggest that it was a personal source of trauma for any of these girls, and it ignored the clustering of the girls' illness. The fresh concerns in the town about the vaccine seemed much more pertinent. Thankfully, since a peace process in 2012, things have improved a great deal in Colombia, and, while there are signs that the peace might not last, it still held when I visited in 2019.

'Everything that was said was disrespectful,' Laura's mother told me. 'They blamed inbreeding, poor diet, lack of exercise. My daughter played volleyball every day. She had a healthy diet. The most awful part was that everybody was mocking us, saying the girls were crazy and they weren't really ill. The national and international press said they were making it up. That was worse than the disease.'

'They said we were uneducated and ignorant,' Laura added.

I was told that some girls were afraid to get better because, if they did, people might say it was proof they hadn't ever really been sick. I looked again at Laura's medical notes and blood-test results and the photograph of her in the intensive-care unit, hooked up to machines. She had been very ill indeed.

I braced myself before broaching what seemed to be the obvious truth. 'I'm so glad you're better, Laura, and I can see how dangerously ill you were. But . . .' I hesitated. 'I'll be honest, I think the vaccine is unlikely to have been the cause for your lupus.'

'She was never sick. She got the vaccine and then she got

sick. That's proof, as far as I'm concerned,' her mother countered quickly, used to defending her view.

'And I'm so glad she's better,' I repeated.

'That was God's will,' Kim broke in again. 'You know what some people said, if the girls went to the hospital? "There's those hysterical women again. They are not getting enough sex."'

It was not difficult to understand why the families preferred the HPV vaccine hypothesis.

Erika, who facilitated my trip, had formed a support group for the community, calling it the Great Feminists of the Maria Mountains. She was certain that the girls' experiences had been easier to dismiss because they were girls and because many were poor. I was sure she was right.

It is inescapable that hysteria is a feminist issue. The word 'hystera' comes from the Greek for 'womb', so there is good reason to have issue with the label. It is not only women who are affected by functional or psychosomatic disorders; men are also affected. Freud had male 'hysterics' under his care, but in *Studies on Hysteria* he only wrote about his female patients. Similarly, all Charcot's best-known patients are women. It is simply incorrect to portray these disorders as being exclusively female territory. However, it is still true to say that these conditions affect more women than men. That's not only the case with mass outbreaks; in the general population, at least two thirds of people with functional neurological disorders are women.

Some people have tried to explain the female preponderance by saying that doctors are more likely to 'dismiss' women's symptoms as psychological and are less inclined to pursue an alternative explanation, thus implying it is a diagnosis of neglect rather than a significant medical disorder in itself. I certainly agree anecdotally that women are more likely to be dismissed by doctors as 'complainers', and that men are more willing to diagnose functional disorders in women than men. Medical communities clearly need to address these issues, but it is the way the diagnosis is portrayed as a 'dismissal' that I think illustrates the real sexism at play here.

There is absolutely no doubt that psychosomatic disorders are more common in women, even allowing for bias. I think that is precisely why the medical community has found the disorder so easy to neglect. For centuries, women's place in society was such that it made it easy for their complaints to be trivialized or dismissed. Attitudes like that still linger today. In the same way that careers, sports and pastimes that are favoured by women are less respected and less well rewarded than their male equivalents, this illness has been regarded as relatively unimportant. Doctors make jokes about women who come to hospital because of dissociative seizures. They are referred to as time-wasters. If it were a disorder that stopped middle-aged, middle-class men in their tracks, it might have attracted a different response. It is not that the disorder is a judgemental label given to women – it is that, as a predominantly female condition, it was easy for a predominantly male profession to think it trifling.

It is worth comparing the way the El Carmen girls were discussed with how people talked about the sonic-weapon victims in Cuba. In El Carmen, the girls were accused either of being sexually frustrated or the subjects of abuse, then they were told that they needed husbands. It was suggested that they had been damaged by violence that had taken place before some of them were born.

Of course, the Cuban cohort were not in poverty and they were protected by their diplomatic status, but, even allowing for that, the tone of the discussion around them was very different.

How sexually satisfied were they? Nobody would have dreamed of asking such an inappropriate question. Were they married? Did they have children? Did they come from single-parent homes? Were any in debt? Were they, or their parents, involved in and traumatized by one of the USA's wars? As diplomats, had any ever been stationed in dangerous places? These details were never mentioned, either in news reports or in the medical papers published about them. If these questions had been asked, many stresses would have been uncovered. Nobody's life is blameless, and they could not have escaped the scrutiny the girls received, because nobody could. But they didn't have to withstand that scrutiny. Half of the embassy victims were men and their average age was forty-three. They were middle-class and educated. Their marital status, family backgrounds and social histories were not subjected to public speculation. And, with so many of them being male and all being educated, powerful people were encouraged to dismiss any suggestion of mass psychogenic illness. Their sex lives and relationship status never arose in conversation because these issues are talked about very differently when it comes to men and women.

Truthfully, nobody really knows why young women are more likely to be affected by these disorders. There are many factors, but I am convinced that their voiceless position in society is one of them. There is a strange, impossible place that women are supposed to occupy, which values a gentle, fragrant femininity that is far too quiet to be natural. The girls of the Miskito Coast were expected to live traditional, conservative lives, while also finding themselves sexualized by older men. Their choices in life were limited. Young women everywhere are told they are equal, but are held back when they try to assert that equality. Colombia has strong legislation that has attempted to give women equality, but what happens in practice is often very different.

There may also be something in young women's physiology that makes them more vulnerable to functional disorders, which is nothing to do with stress, psychology or society. The frequent bodily changes that come with cyclical hormones might create more abundant white noise, which young women have to learn how to decipher. A tendency to lower blood pressure and thus to fainting would certainly act as a trigger for dissociative seizures. Women may be physiologically more vulnerable in a healthcare system with a tendency to view the physiological differences of women as weakness rather than biology. The fainting and swooning of women in the 1800s probably had more to do with tight corsets, inactivity and innate low blood pressure than with psychological frailty, as was sometimes implied. Similar judgements still exist when it comes to period pains, the menopause and other female-only medical problems.

I did not meet Erika, the founding member of the Great Feminists of the Maria Mountains, in person until the final day of my visit. While planning the trip, I knew her only through a series of emails and WhatsApp messages. She had often been vague about her exact relationship with the girls. Catalina and I had speculated a great deal about her agenda and identity, but assumed it would be revealed and verified when we finally met her. I had expected her to accompany me as a chaperone for the families. Instead, on my first day in El Carmen, she had sent me an odd picture message which showed the top of her head poking above a duvet, with her face hidden. She told us she was sick, and that was why she had sent Carlos in her place. This really heightened my curiosity and suspicion, and it came as a relief when, on the very last day in El Carmen, she invited me to her home.

El Carmen is set out on a grid around a central square, on

which there is an oversized church. The streets are busy with people and motorbikes. The houses, with their open doors and windows, look airy and bright, albeit often in need of some renovation. Erika's front door was not one of those that hung open all day. It was heavy and wooden and firmly locked against the world. Her house was significantly bigger and grander than any of the others I had visited, although it was also slightly run-down. The rooms were arranged around a courtyard filled with plants and with a water feature in one corner.

Erika, a small, middle-aged woman, greeted Catalina and I warmly on our arrival and ushered us to a leather sofa. I had been nervous about the meeting, but there was none of the oddness I had expected after receiving her strange picture message. The negotiations to arrange the trip had been tense. Erika was very concerned about the way the girls had been portrayed and expressed strong objections to the mass hysteria label. She had requested that I not mention that diagnosis in anything I wrote, and I feared the arrangement would fall through when I said it would have to feature in the discussion. In the end, she surprised me by facilitating the meetings despite her reservations. Now, seated in her home at last, I found her pleasant and likeable. We got to know one another over tea and biscuits, while her husband pottered in the background, watering plants. For once, Carlos was not with us.

When we had finished with small talk, Erika brought us to her office. She sat behind a big wooden desk, surrounded by piles of paper and computers. Catalina nudged me and pointed to security cameras in the corner of the room, and I later noticed them elsewhere, inside and outside the house. We realized, over the course of the conversation, that Erika was actually a recluse, who feared for her safety because she took a vocal stance against government corruption. I now suspect she never intended to escort us to meet the girls, and that the sickness was a ruse, although I may be wrong.

The move to the office indicated that we could talk business. Until then, we had been skirting politely around the issues. It was clear that Carlos had kept Erika fully informed about my discussions with the young women, and she immediately expressed concern that I would refer to their problem as psychogenic – that I would paint them as crazy and broken. But, of course, I did not think the girls' problems were an issue of their individual psychological states at all. I did not find them to be weak or troubled. The outbreak in El Carmen was being driven forward by many people, least of whom were the girls themselves. To my mind, neither the cause nor the solution lay with them. A simple mass fainting episode had been turned into a protracted medical and social problem by fear-mongering and disinformation. It had also been stoked by poor communication of the diagnosis. They had been given the correct diagnosis, but had not understood it. How could the girls get better if they believed they had been poisoned? How could they accept a functional cause for their symptoms if that was taken to mean they were crazy? I didn't think the girls had psychological problems. It seemed to me that the solution lay with the parents and the community, not the girls.

'It is simply not possible for a girl to have a seizure for five hours for psychogenic reasons,' Erika said.

But, of course, that is not only possible, it's common. In fact, epileptic seizures are generally brief, while dissociative seizures last a long time. As usual, the misperception that functional symptoms are always mild and self-limiting had encouraged people to doubt the diagnosis. I wondered what she would think of the resignation-syndrome children's plight, some of whom had been comatose or catatonic for years.

Through Catalina, Erika and I had a protracted conversation about the modern interpretation of mass hysteria. I agreed the label 'mass psychogenic illness' was inadequate and misleading, but, problematic as it was, it was still the correct diagnosis. I discussed the very real nature of functional symptoms, how they come about and how looping can make symptoms accumulate, thus sustaining disability for a long time. To my relief, Erika nodded thoughtfully and listened carefully as I spoke.

'When you put it that way, I can understand that you do not mean the term "mass psychogenic illness" pejoratively,' she said.

I had not expected her to take the conversation so well, since it had been such a sticking point during the planning of my visit.

'I think we need to have this conversation with the girls and the families,' I suggested.

'They won't understand,' she said.

'I think they need to be given the chance.'

'And I don't want you to write that it's mass psychogenic illness in your book,' she said.

My relief was short-lived.

'Not being frank with the families deprives the girls of a chance to get better.'

'If the president doesn't understand, why would your readers?' she said, referring to the president of Colombia's public statement about MPI, which had caused so much distress. She pushed one of the large piles of paper towards me on the desk. 'I would like your opinion of this doctor's work.' Then she turned her computer screen to face me. It displayed a website for a doctor called Juan Guzman. It featured a series of articles about the girls, and some others about more general issues related to Colombia. Some were in Spanish and others in English. The author identified himself as a medical doctor who had graduated in Colombia, although the website was a personal one. He did not mention his area of expertise, nor did he have any links to employers or publications.

'Where does this doctor work? Is he a specialist?' I asked, as I scrolled through the pages.

Although it didn't seem like a very difficult question, it kept Catalina and Erika talking for a few minutes. Eventually, Catalina turned back to me with a grimace.

'This is a little odd, but Juan Guzman isn't his actual name. She doesn't know his name. She says he's a Colombian doctor exiled in the USA, who can't reveal his real identity because he fears for his safety.'

I tried not to show how horrified that made me, and asked Catalina, 'How does she know he's a doctor?'

In response, Erika indicated the webpage, which certainly looked superficially professional, but which was not linked to any academic institution. I started leafing through the paperwork she had given me. It mostly comprised copies of medical questionnaires which many of the sick girls and their families had completed. Erika informed me they were part of a research study and Juan had used them to diagnose the girls with a rare autoimmune disorder. He had never met the girls, but had held video conferences with some of them.

'He has video conferences with the girls? I assume their parents sit in?'

Erika didn't know. She facilitated the meetings, but didn't supervise them.

When I looked closer, the questionnaires turned out to be standardized pain scales. They required the user to rate their experience of pain and indicate on a picture where the pain was located. Medical illustrations are routinely made on non-gendered outlines of a person's form, but these seemed to have been personalized by Juan, because each featured a picture of a young girl with long black hair and a big smile, wearing knee-high white socks and a pinafore. I had never seen a schoolgirl's picture used in this way in a medical setting before. It made my skin crawl.

'Who is Dr Herrera?' I asked, immediately thinking again of the other doctors and researchers I had been told about.

'I don't know Dr Herrera. I have only spoken to him once. He is Carlos' friend.'

She knew a little more about Mila, who, it transpired, was a doctoral student based in the Netherlands, although Erika could not speak to the nature of her thesis or immediately advise me of her full name or her employer.

'You say the problem is mass psychogenic illness,' Erika said, 'but can I ask you to look at Dr Guzman's work and tell me if he might be right?' She indicated Juan's website once again. The screen page she had stopped on was littered with the acronym HPV. As receptive as Erika had been to our discussion about functional symptoms, it seemed she was preparing to dismiss it again very quickly in favour of Juan Guzman's more palatable diagnosis. Juan's website was very detailed, and in two languages, so I promised to look at it and respond to her once I was home.

As a heavy tropical rain began to pelt the streets outside, Catalina and I spent the rest of the afternoon talking to Erika about the sort of background checks required before a person can work with children. We discussed Internet fraud and online security. Catalina and I offered to help verify the identity of any future researchers coming to the town. Erika nodded and listened carefully, just as she had during the discussion about functional disorders.

I told her that, in Ireland, where I am from, a lot of people once had similar concerns about the HPV vaccine, but that the incredible activism of a young woman with cervical cancer, Laura Brennan, had taught us that the vaccine was both safe and necessary. Women in El Carmen are sexually active and have children at a younger age than in many countries, which puts them at a higher risk of cervical cancer. I wondered where all the naysayers against vaccination programmes would be when the consequences of the fall-off in vaccination rates started to show themselves. Erika continued to nod and look attentive as we talked; in the end, I still wasn't sure whether we had found some common ground or had just agreed to differ. We said we would correspond again once I had read Juan's work in more detail.

Back out in the street, with the heavy wooden door closed behind us, I asked Catalina, 'Do you think I got through to her?'

'I think you did.'

The rainfall had left the roads waterlogged, and we waded back to our hotel through the rivers of mud Frida had promised us on our first night.

The El Carmen story has not yet had its denouement and, as things stand, there is no end in sight. Even in 2019, there were further outbreaks of seizures in schools hitherto unaffected.

The people of La Cansona gathered in a semicircle under a corrugated roof in the Maria Mountains were my last meeting in Colombia. It became obvious very quickly that I had been mistaken when I thought they were angry; in fact, they were worried about their children and desperate for help. In comparison to the affected families in central El Carmen, they had difficulty accessing medical care. La Cansona was beautiful, but poor. Frida and Yesmid and the other girls might have complained about the help they received in the hospital in El Carmen, but these people didn't even have the means to make regular trips there. 'La Cansona,' I learned, means 'the tired lady,' and the name is said to refer to how one feels at the end of the gruelling journey to the top of the mountain. When the sickness hit the schools in the region, the parents had to carry their children on borrowed motorbikes down the hill into El Carmen. Some girls fainted, they told me, others convulsed, others couldn't walk. A few had hallucinations, plucking at invisible things in the air, just like the children in Krasnogorsk. They screamed as if they were having nightmares. They grabbed at their own necks, as if trying to hurt themselves. It took four grown men to restrain a girl who was having a seizure, I was told, in echoes of grisi siknis and my own patients' seizures.

'It was the vaccine,' the man in the red shirt had told me.

His daughter, Maria, was with him, and I asked her to tell me about the experience.

'Everybody was talking about the vaccine,' she said. 'They said that, because I'd had it, I would never be able to have children.'

'Was this before or after you got sick?'

'Before,' she answered. 'Some parents were told the vaccine would kill their children. When I eventually got sick, I felt very lonely and very scared.'

Maria had collapsed during a wave of fainting attacks in her class. I asked her why she thought all the girls had collapsed together.

'Nerves,' she said. 'One collapsed and then another. I fainted because I was scared and nervous.'

I looked at the gathered people, but I could see no sign that they appreciated the wisdom and insight that Maria was offering, any more than Jenny had appreciated Frida's thoughts about the cause of the initial seizures at her school.

Maria's father sold his cows to pay for herbal remedies, and he attributed her recovery to that.

'I would have sold myself, but nobody would buy me,' he said.

We laughed, but it was also very sad.

'We've talked to so many people. So many people have come,' another parent told me, referring again to the many outsiders who had come to the area at the peak of the outbreak. Journalists wrote about them, but didn't come back, they told me. Doctors had come from other countries to tell them Big Pharma and the government were misleading them. Researchers said their daughters had been poisoned by aluminium in the vaccination.

I cannot say how the original connection between the vaccination programme and the seizures arose. Many parents reported that their children were vaccinated without their consent, so there was plenty of room for conspiracy theories to grow. Then there is the batch of vaccine that was, according to Frida and many others, mishandled and contaminated, which I am sure is also pertinent. What is very clear, however, is how the anti-vaccination message took such unshakeable hold. The opportunists, activists and non-specialist scientists who contacted and visited the town after they learned of the outbreak reinforced the fear of the vaccine with liberal misinformation, until even the girls who were aware of the connection between their own fear and the seizures dismissed this important link. El Carmen was once isolated from the rest of the world by the violence in the region. They lacked for foreign visitors. Even now, not every person living in the region has easy access to the Internet. It hadn't been difficult for foreigners with the prefix 'doctor' before their names to create an impression of authority. Activism that promised to help the girls was actually harming them. I believed that Erika's intentions were good, but, to my mind, she was in fact harming the girls without realizing it.

The voices of young women are often not respected; they are

the group most vulnerable to conjecture and insulting speculation. It had been far too easy for a lot of strangers to exploit the girls' circumstances for their own gain. Children and young women were now trying to recover in an atmosphere in which they were being told that recovery was impossible.

Neither were the parents being heard. Every time I asked them what they wanted, they made the same crystal-clear requests. At the top of the Maria Mountains, I asked the gathered group what they wanted me to say on their behalf.

'We need psychological support,' the man in the red shirt told me. 'We, the parents. And we need financial help. We have sold all our belongings to pay hospital bills.'

I looked from person to person, and every one of them nodded in agreement.

I had to wait until I was home to research the various visitors who had been advising the people of El Carmen.

Mila, who seemed to have helped Frida a great deal, did not work for Médicins Sans Frontières. She was a Netherlands-based PhD student, with a Master of Arts degree in development studies. She does not work as a psychologist and she does not treat patients. Her thesis was on the embodied experiences of HPV vaccination.

Once I had translated and read Juan's articles concerning the girls, I found them worryingly unscientific. None were peer reviewed, so, rather than being medical papers, they were effectively little more than self-published opinion pieces. Having read his work, I would say it is unlikely that Juan is a medically qualified doctor, and he certainly has no knowledge of neurology. Erika had placed a great deal of trust in a diagnosis of an autoimmune disorder, which Juan had made through his questionnaires and online conversations with the girls. That diagnosis was far-fetched, to say the least.

As Carlos had promised, I was able to watch Dr Herrera's lecture to the crying 'doctors' in the USA online. It was not a medical conference and many of the group were not doctors. The meeting was an anti-vaccination platform. Dr Herrera was identified as an anthropologist and journalist, and I could find no affiliations to any professional organizations for him. In his lecture, Dr Herrera told how, in 2012, two years before the El Carmen girls started to have seizures, he wrote a blog post for one of Colombia's biggest newspapers, *El Tiempo*, in which he claimed that the HPV vaccine was killing girls. *El Tiempo* had apparently taken down the blog and instructed him to print a retraction. He proudly told the delighted conference audience that he had written a fifty-page follow-up reasserting his claim, but *El Tiempo* did not publish that.

Just before I left El Carmen, Carlos told me that Dr Herrera was scheduled to return to the town with three Polish men who he said had invented 'a machine' that could cure the girls. I strongly advised Erika to put a stop to the visit, and she said she would look into it. For a brief period after returning home, I was privy to a series of WhatsApp messages as part of a group set up by Erika. It was clear from the exchanges in that group – which included some from Juan – that my intervention had not shaken her trust in Juan or Dr Herrera, and had not dented her certainty that the HPV vaccine had caused the seizures.

After some sleepless nights, I reported the situation in El Carmen to ECPAT, an organization that works to end online exploitation of children, and also to Save the Children, who passed the report on to international law enforcement colleagues.

7

The Witches of Le Roy

Mass Hysteria: *Characterized by excitement or anxiety, irrational behaviour or beliefs, or inexplicable symptoms of illness affecting a group.*

Public interest is easily piqued by reports of so-called mass hysteria. Accounts of it are often presented as something of a freak show, full of histrionic, fainting schoolgirls and epidemics of bizarre behaviour. Truths about the condition are misunderstood and misrepresented, both by the media and by less well-informed members of the medical community, of whom there are many. Remember how mass psychogenic illness was conflated with malingering by senior medical experts investigating the cause of Havana syndrome. Meanwhile, the El Carmen community assumed that MPI was something that only happened to damaged people.

More than most medical problems, this one seems unable to escape being weighed down by cliché and long-ago debunked theories. A particular issue is that the disorder is inevitably seen as arising from inside the person as a result of some fragility belonging to them, when actually that is not typically the case. In fact, a true mass-hysteria outbreak says much more about the society in which it occurs than it does about the individuals affected.

Not long before the events of El Carmen began to unfold, something very similar happened in two very different communities, one in the USA and one in Guyana. While the demographics of the affected groups were very similar – teenage girls in small-town schools – the social driving forces behind their experiences had little in common.

The first of these took place in 2011, in Le Roy, New York. Lying 350 miles north of Manhattan and seventy miles east of Niagara Falls, Le Roy High School was the site of a very widely publicized epidemic of neurological symptoms, labelled, by many media outlets, as mass hysteria.

It is rumoured to have started with high-school senior, Katie Krautwurst – a straight-A student and cheerleader, with lots of friends. I first saw Katie pictured in the *New York Times Magazine* and, as wildly different as the women were, she made me think of Lyubov. Like her Kazakh counterpart, Katie was shown staring into the distance, wearing a forlorn expression, but there were hints, too, that this might be a young woman with a flamboyant side. She wore joyously mismatched multicoloured socks and her pink bedroom was littered with girlish trinkets. I wondered if the photographers who had taken these pictures of Katie and Lyubov had instructed them not to smile, or if the two women genuinely felt as despairing as their photographs made them look.

Katie's illness, I learned, began in October 2011, when she awoke from a nap to find she had developed involuntary movements and verbal outbursts similar to those seen in Tourette's syndrome. Her jaw went into spasm and her face twisted. She twitched and writhed and let out involuntary shouts. It took a couple of weeks for the contagious element of Katie's symptoms to reveal themselves. Her close friend and fellow cheerleader, Thera, was next. She developed almost identical motor and verbal tics; she stuttered and her limbs flailed. After that, the disorder moved to others in the girls' close circle, and then later to a wider group within the school. Faithful to Hacking's description of the classification and looping effects, as it spread, it changed. In new victims, the disorder became more flagrant and disabling. Some of the girls had full convulsions, in keeping with dissociative seizures. Some couldn't walk. Over time, the symptoms suffered by both Katie and Thera also evolved, with each girl ultimately needing a wheelchair when the muscle jerks became so violent that they caused them to fall.

The North American schoolgirls were more fortunate, in many regards, than those in El Carmen, not least because they all had access to a large centre for neurology. Ten of the original twelve were seen by the same neurology team, in Buffalo. When there were only a couple of victims and the connection between them was not immediately obvious, a diagnosis of Tourette's syndrome was considered. Once it became clear that the tics were contagious, and the condition began to evolve and spread among a friendship group, the diagnosis of Tourette's syndrome became untenable. The girls were fully investigated and ultimately diagnosed with a functional neurological disorder under the name 'conversion disorder'. The school took the outbreak very seriously. They arranged for environmental tests to be done; the Centres for Disease Control and Prevention were consulted and the New York State Department of Health also became involved. Environmental toxins and infectious agents were ruled out.

For the first three months after the outbreak started, things seemed to be contained. The affected families put a reasonable degree of trust in their doctors and accepted the reassurance they were given. However, sadly for everyone involved, that changed when, in January 2012, a meeting was held at the school where the Department of Health announced the results of their inquiry to a wider audience. At that time, only the affected girls and their families knew the full details of the diagnosis they had been given. The Department of Health assured the gathered parents and students that the school was safe. They also vaguely alluded to the fact that there might be a stress-induced cause for the outbreak, although they did not go into details, citing privacy laws. There were aspects of the case they could not discuss at an open meeting.

The stress hypothesis transpired to be inflammatory – it made no sense to many in the group, and the lack of familiarity with the common nature of functional symptoms meant nobody really believed that psychological processes could produce the sort of neurological complaint seen in the girls. That information seemed to be being withheld created an atmosphere of mistrust. It is in the gap between the diagnosis of conversion disorder (functional neurological disorder) and the understanding of what that diagnosis means that there is space for harmful things to grow. In El Carmen, the 'mass psychogenic illness' label created confusion, and, in Le Roy, the diagnosis of conversion disorder did the same. This is common in everyday clinical practice, too; a functional or psychosomatic diagnosis is often misunderstood. Most of what people think they know about these conditions is wrong. When they rail against a functional diagnosis, they are actually rejecting a long-since debunked interpretation of the condition, not the disorder as it is understood now.

The sceptical voices of the wider group began to play on the minds of the affected families. And, once doubt about the validity of the conversion-disorder diagnosis had spread, they were keen to find a better answer. With the school and doctors determinedly supporting that single explanation, some parents felt they had no choice but to seek new avenues of help. Media exposure is often used as a way of pressurizing others into action, and, with that in mind, one of the parents contacted a journalist. They cannot have foreseen the effect this would have on the town.

The story of what was happening in Le Roy came to widespread public attention in mid-January 2012, when Katie, Thera and their mothers appeared on NBC's *Today Show*. Sitting on the faun breakfast sofa, Thera twitched nervously and insisted that her life had been stress-free, right up to the day her symptoms began. Katie seemed better able to control her tics – that day, at least – and sat with her shoulders slumped, looking young and vulnerable. The television appearance was a plea for action and for answers. On live television, they fought against a psychosomatic explanation and expressed their distrust of the school and the doctors. And, foreshadowing what would soon happen in El Carmen, the group complained that they had not been given all the test results – that people were withholding the truth. These were popular, well-adjusted children, the mothers argued. They demanded more tests on the girls, and on the school environment.

The interviewer, aware of the conversion-disorder diagnosis, put it to the girls and their mothers.

'I'm very angry,' Thera told the interviewer, and her mother added, 'The facts that they're stating just are not true.'

The Today Show had a powerful effect. For a week, the girls were the lead story on multiple news stations throughout the US. They also caught the attention of the world's media, with headline stories like this, from the *Daily Mail*: *Mystery illness gives 12 girls at same school Tourette's-like symptoms of tics and verbal outbursts.* Here was a human-interest story with the added attraction of a potential conspiracy. Why had the doctors dismissed the girls so quickly? What was the school hiding? The media was attracted by the voyeuristic controversy that comes with any suggestion of mass psychogenic illness. *Mass hysteria outbreak reported in NY town,* CBS News said, attributing the problem to 'psychological conflict'. In explaining what was meant by this diagnosis, the CBS report continued, 'For example, a woman who believes it's not acceptable to have angry feelings may experience numbness when they get really mad.' Only a woman?

Nothing the school or the State Department said by way of reassurance seemed to dampen the media's enthusiasm to stir up the *mystery* element of the narrative. The doctors were incorrectly referred to as 'stumped'. Le Roy became a worried, anxious place. Concerned parents demanded answers. People were scared to send their children to school. New cases emerged, among them a boy at the school and a middle-aged woman unconnected to it. The day after the NBC interview, two of the doctors involved in the case were given permission to speak to the media about the diagnosis. In statements on local news programmes and to NBC *Today,* they carefully explained the conversion-disorder diagnosis. Both took great pains to emphasize the unconscious nature of the symptoms and the reality of the girls' suffering, albeit with liberal use of the word 'stress'. Undeterred, news reports continued to express concern that the doctors had dismissed the girls too quickly.

While the media attention appeared to put the families back in the driving seat, ultimately it exacerbated the situation: the girls became a spectacle. Over and over, they were shown on news programmes and on social-media platforms twitching, stuttering and convulsing. The cameras pointed at them shamelessly, as they struggled to speak and sit still. When the girls were challenged with the conversion-disorder diagnosis, they replied, 'Why would we fake this?'

From there, the media speculation had two main camps – those who accepted conversion disorder as an explanation and those who did not. The former had hackneyed discussions that conflated the condition with a hundred-year-old version of hysteria. 'Stress' became the byword for what was happening. Parents were forced to defend their children in public – their daughters were not traumatized, not mad, not pretending. The media picked over the girls' lives to find evidence to prove otherwise. They listed the factors they believed were the cause: parental illness, single parenthood, poverty, family disputes. It had proved easy to find hardship in the girls' personal lives on which to hang the blame for their illness, ignoring the fact that everyone has some skeleton in the closet or source of unhappiness waiting to be found. That Le Roy was a town long past its heyday didn't help.

Le Roy had been made fortunate by industry, and the people who lived there had once been above averagely blessed. Aside from being the scene of a mystery illness, Le Roy was also the birthplace of Jell-O – or jelly, as it was known in my Irish childhood. Jell-O had made the town wealthy. For the first half of the twentieth century, the grand houses were freshly painted and the main street bustled. But, in 1964, the Jell-O factory was relocated to a larger town, which came as a huge blow to Le Roy. The loss of this pivotal source of employment meant a significant reversal of fortunes for the town. The history of the Jell-O factory and the town's fall from prosperity became the backdrop to how the media presented Le Roy's brush with so-called 'mass hysteria.' It was not dissimilar to the bleak way in which the story of Krasnogorsk had been told. But there was even worse in store for the people of Le Roy than what their older counterparts had suffered in Kazakhstan. As almost all of those affected were young women, it was perhaps inevitable that witch trials would be mentioned, 300 years after the fact. An article in the *American Spectator* was headlined, *The Witches of Le Roy*.

Not every media commentator agreed with the conversion-disorder diagnosis, however; a second camp took the opposite stance, regarding the diagnosis with suspicion and incredulity. Just as would happen years later in response to the Havana-syndrome outbreak, conversion disorder became 'just' a conversion disorder, and hysteria, 'just' hysteria. Doubters claimed the doctors hadn't looked hard enough for an alternative answer. They called for more tests and suggested the doctors and school authorities were not doing their jobs.

The public made it their responsibility to uncover the cause and reveal the conspiracy. In a classic case of recall bias, many amateur investigators sought to uncover anything odd that had ever happened in the region of the school, and, as Le Roy is an old industrial town, there was plenty to draw on. Somebody told reporters that, when the Jell-O factory was open, the creek that ran through the town used to change colour according to the flavour being manufactured on any given day. Who knew what toxins had been left behind? Somebody else remembered a barrel in the river where the children used to swim. Once, a yellow-orange ooze had been seen on the athletic field where the girls were cheerleaders. Crop dusting and fracking were also mentioned among the many fears, while familiar culprits like vaccinations were considered, briefly, but, fortunately for Le Roy, were dismissed before they could take hold. A variety of medical explanations were suggested by outside sources, and some were seriously entertained, with a selection of the girls being diagnosed with PANDAS, a rare autoimmune disorder believed to be triggered by a streptococcal infection, although most doctors, including the expert who originally described PANDAS, dismissed it as impossible.

Of all the causes suggested, one took hold more than any other, creating a frenzy of anxiety and excitement that was hard to quash. It began when somebody slipped a note into one family's mailbox, which advised them that a train crash had left a toxic spill a mere four miles from the school. The worried mother who received the note took the unconventional route of contacting celebrity investigator Erin Brockovich, who quickly took the bait. Brockovich came to public attention for her dogged investigation into the poisoning of Hinkley, another small US town, by contaminants from a Pacific Gas and Electric plant. She garnered worldwide fame when her story was made into a major movie, in which she was portrayed by Julia Roberts. When Brockovich became embroiled in the Le Roy case, her fame and previous success threatened to completely overshadow the doctors' statements. The tone of her involvement was captured by a Reuters reporter, who described how she had 'scoffed at the psychological diagnosis'.

In early 2012, Brockovich appeared on ABC News in an interview with celebrity doctor, Drew Pinsky, to discuss events in Le Roy. At this point, neither she, nor anyone affiliated to her, had conducted a single test in Le Roy and she had no privileged information. That did not stop her from having strong opinions on the case.

Brockovich made her thoughts clear in her opening exchange with Pinsky, when she expressed concern that news of the incident had hit the media very quickly, and that the diagnosis had been far too sudden for her liking. (The implication being that all medical diagnoses are difficult and that an easy diagnosis must be wrong. The opposite is true. Most experienced doctors can make a reliable diagnosis within a few minutes of meeting a patient – but media portrayals of medicine show the conundrums, making the undiagnosable cases seem much more common than they are. It wouldn't make for very good television if hospital dramas got it right too easily, and the people who have common medical problems don't usually make it into the newspaper.) The certainty of the specialists' diagnosis of the Le Roy outbreak did not have much impact on Brockovich or Pinsky.

'Did you have the same reaction as I did?' Pinsky asked. 'The diagnosis seemed . . . it didn't feel right in your gut, and then it kind of closed the door on further investigation.'

The diagnosis felt 'too convenient', Pinsky told Brockovich – although convenient to whom he did not say. Pinsky repeated the hackneyed debates already being played out in the press: the belief that the girls just hadn't been tested enough and that, if the doctors looked hard enough, they would find the real diagnosis. Brockovich agreed.

They focused heavily on the train crash. Brockovich had not been to Le Roy, but she had done an Internet search. The crash was real and there had been a significant spillage of cyanide and other chemicals on the site. However, it had happened in 1971. During their conversation, the pair sat in front of a map of Le Roy. The high school was highlighted at the bottom left corner, with the site of the train crash, not far away, at the top right. Pinsky eventually asked the obvious question: 'How could a crash that took place in 1971 affect the school forty years later?' Brockovich reported that, in 1999, a heavy rain was known to have caused a plume of toxins to move through the ground. The plume had travelled north-east for at least a mile. Once again, Pinsky was duly horrified. Neither seemed to notice that the crash lay north-east of the school, which meant that a 'plume' travelling north-east was going away from the school, not towards it.

'I am going there today,' Brockovich told Pinsky.

'I'll send some of my team with you, if you don't mind?' he asked, and she agreed.

They went on to talk about the types of frightening chemicals that might be involved, all with the potential to be neurotoxins. Halfway through the segment, Pinsky paused to acknowledge that it was still necessary to link the toxin leak with the biology causing the girls' symptoms. He also expressed concern that their very conversation could contribute to panic in the town. He pondered aloud how they could try to minimize that panic, before adding, 'Mind you, I'm feeling panicky just talking about it, and I'm not the one living under a plume of TCE!'

Over the course of a short conversation, a forty-year-old underground contamination of uncertain significance, which lay four miles from the school, had become a cloud of TCE covering the town. It was visually very compelling, a little like Tara's mental picture of a slipped disc cutting through her spinal cord, or the idea that sound causes brain damage because it enters through the ears. Every acknowledgement of the lack of actual evidence was immediately overshadowed by an inflammatory, excited sidebar.

'I want evidence that connects the dots biologically, here . . . this [toxin spill might] turn out not to be the cause of the tic thing,' Pinsky said, followed by a quick about-turn. 'But I bet, if we examined the cemeteries in Le Roy, you'd see a lot of cancers, a lot of infant deaths – way more than average.'

Why only terrify a school full of teenagers, when you can terrify a whole town?

Brockovich's team did visit the school that day, as did Pinsky's news team. Numerous other local, national and international press were also there to record the event. Unfortunately, it did not go well. The school superintendent, Kim Cox, and the local police were waiting. The area had already been tested under the jurisdiction of the local authorities. Brockovich's team were not authorized to enter school grounds and were escorted away. An angry debate followed, which saw school officials cast as the bad guys – after all, nothing fuels a conspiracy theory more than plain and frank denial of it. As one of Brockovich's investigation team said, 'When I'm confronted by officials barring access to something, they usually have something to hide.' That, or they had schoolchildren to protect.

All of this only served to increase tensions within the town. Another community meeting was called, at which a panel, led by Kim Cox, faced off against parents, the media and other townspeople. It got very heated. Cox reassured the group that the school building was safe and the children were not in danger, but no one believed her. Parents demanded to know what was being done to protect their children.

'You're not doing your job,' one mother shouted at the panel, to enthusiastic applause. The crowd demanded answers, as if nothing had been done and nobody had told them anything.

This was not so different to El Carmen. In both places, unqualified outsiders were given more respect than medical teams. Cox listed the environmental tests that had been carried out, all of which were clear, but it seemed to bounce off the worried, fired-up audience, which mainly consisted of frightened parents desperately trying to figure out how to protect their children. Many walked out in disgust. It was almost as if the only way to satisfy the people was to confirm their fears, because denial only seemed to imply either a cover-up or a blunder. In the end, the panel could only promise that they would arrange further tests, carried out by an independent body.

For weeks afterwards, news reports repeatedly referred to the doctors as 'baffled'. Even in articles that mentioned the conversion-disorder diagnosis, the focus was still on the mystery. While carrying out their own tests, Brockovich's team continued to assert that spilled chemicals could be causing the girls' symptoms, but they never addressed or appeared to worry about the biology. How could an old chemical spill cause such acute tics over a period of only a few weeks, and only among the girls? What part of the brain was affected to produce their odd constellation of evolving symptoms that made no anatomical sense? Brockovich speculated that the school might have been built on soil taken from the spill site, but that still didn't explain the selectivity of the outbreak, not to mention answer the question of why the school or the Department of Health would lie to the families.

The press furore around Le Roy lasted weeks, but in the end this town and these girls turned out to be much luckier than the people of El Carmen, because their story did have an ending. After an intense period of escalating symptoms, the total number of victims stopped at around eighteen, following which the severity of the tics and convulsions began to recede. However, the whole event had gone on for far longer than mass hysteria outbreaks usually do, and the reason for that, and for the ultimate resolution, was, to my mind, sociocultural, not personal. The media frenzy, the misrepresentation of conversion disorder, the public stigma associated with biopsychosocial illness and, more than anything else, the atmosphere in which opinion was given the same weighting as fact – all these factors had stoked the worst of the hysteria in Le Roy. And yet, the story of the town was steadfastly presented as a psychological problem belonging to the girls, as if all those outside elements didn't exist. Only passing recognition was given to the reality of what a media- and celebrity-driven culture might have done to them. Ultimately, it was the withdrawal of the families from the many malign external influences, and the doctors' strong stance in presenting conversion disorder as a valid and positive diagnosis, that brought the outbreak to a close.

Let me compare the story of Le Roy to one that took place in Guyana a couple of years later. This was another high-school outbreak, referred to as mass hysteria. On the surface, these two events seem to have something in common, but it is the way in which they are different that shows the folly of blaming MPI outbreaks on the psychology of young women – or on any individual, for that matter – while failing to see the bigger picture.

I doubt anybody in Le Roy or El Carmen has heard of a small town in Guyana called Sand Creek. Why would they have, when it is situated in a remote jungle area and has a population of less than a thousand? But, shortly after Katie and Thera developed Tourette's-like tics, and not long before Frida and Juliet and their classmates got sick, the young women of Sand Creek high school were in the grip of a very similar health crisis. The story of how it unfolded was told to me by Courtney Stafford-Walter, a young American anthropologist who happened upon the phenomenon in 2015.

Courtney was completely unaware of 'the sickness' when she arranged to spend a year in Sand Creek. The purpose of her visit was to study the effect of an expanded education programme rolling out across remoter parts of Guyana.

Sand Creek is a tropical place, nestled between jungle-covered mountains. At certain times of year, when the climate is right, the region is beset with swarms of mosquitos. Most people in Guyana live on the coast, but Sand Creek is far inland. Located in Guyana's Region 9 (the country is divided into ten regions), it is both the largest and the least populated area of the country; resources are scarce compared with the coastal areas. It is a farming community, in which the men often have to leave to get work elsewhere. Typically, they work as miners. Some women also leave for work, usually taking domestic roles, but that's less common. It is more usual for women to stay at home, where they take most of the responsibility for matters within the village, while the men liaise between the women and outside communities.

Like Nicaragua and Colombia, the population of Guyana is mostly mixed ethnicity, with only a small indigenous population of about 7 per cent. The Dutch were the first to colonize the region, and the British were next. Most people in Guyana are Indo-Guyanese or Afro-Guyanese, descended from servants and slaves. Guyana is the only South American country in which English is the official language, although many of the indigenous people have managed to retain their own languages. Sand Creek is home to the Wapishana, one of the Amerindian indigenous tribes. They speak English and Arawakan, and, like the Miskito, they are caught between a traditional way of living and modernity.

Because of Region 9's remoteness, there had always been limited schooling available for Amerindian children. The population was sparse and disparate, making it difficult to provide the same education for everyone. Each village had a state primary school, but there was only one secondary school for the entire region and only the brightest pupils were able to progress to that stage of education. The system usually favoured boys. To improve availability of secondary-level education to everyone, the government set up three more state secondary schools in the area. There were practical issues getting to these schools, though, as the roads were poor and there were very long distances between villages. As a result, every school had boarding facilities and, in the recently founded Sand Creek Secondary School, half the students were boarders.

When Courtney moved to the area, Sand Creek Secondary School was understaffed. Although she was not a teacher, she was co-opted into teaching a class, an opportunity she welcomed; it would bring her closer to the children whom she had come to observe, not that it proved to be an easy job. The students showed little respect for her authority, chatting during lessons and coming and going from the classroom as they pleased. Other teachers maintained control with the threat of corporal punishment, but Courtney was reluctant to emulate them.

One morning, Courtney turned up to class to find her usually exuberant, hard-to-corral pupils subdued. When she asked the children what was wrong, they said they were upset because one of their friends was sick and had been sent home. 'Granny took her,' one of the pupils said. Courtney had been aware that a 'sickness' was sweeping through the school. She had heard it referenced many times, usually in vague terms and with hushed voices. In the first instance, she assumed it was a tropical illness, such as malaria. But the more she heard it talked about, the less certain she became. The tone with which 'the sickness' was being discussed made Courtney doubt that 'Granny' was just a family's kindly old matriarch, and it took time for the local people to trust Courtney with the truth.

The first girl to get 'the sickness' in Sand Creek caught it in 2013, a year after the school was founded. The main symptom was seizures. The affected person typically experienced long episodes of lying on the floor, unconscious, limbs flailing, and frothing at the mouth. It was also a highly contagious disorder and it spread through the school in waves. At the worst times, there were new cases every day. The seizures happened most commonly in the dormitories, and there were sometimes as many as half a dozen girls convulsing on and off throughout the night.

When it began, the school responded, as one would, by calling a doctor. The village was still somewhat reliant on shamans to treat familiar minor ailments, but in general they associated Western medicine with advancement. A skilled health worker was stationed in the village, but seizures were beyond their expertise and a doctor had to be called from a larger town. He arranged for the sickest of the girls to be taken to hospital by aeroplane, where they underwent tests, all of which came back normal. Interestingly, the seizures had stopped as soon as the girls left Sand Creek. Like grisi siknis, the illness seemed to be tied to the place. The doctor couldn't provide a coherent explanation, so the local people quickly developed their own theory about what was making the girls sick.

The Wapishana are spiritual people, for whom ghosts, witchcraft and magic exist as part of their everyday life. Beliefs surrounding death and illness differ from those of Western societies – they are associated with agency. One doesn't just fall ill; rather, illness is done to a person by another. That other could be a neighbour, a friend, or it could be someone magical. In the case of 'the sickness', the agent of illness was 'Granny' – a spirit. The school dormitory nestled against a jungle-covered mountain, and the villagers believed that the spirit of an old woman lived in a cave partway up the mountain. They guessed that one of the schoolgirls must have ventured into the cave and disturbed the spirit, which is how Granny had come to possess or haunt the girls.

At first, for Courtney, 'the sickness' remained a rumour. She had heard some descriptions of the seizures and had seen a grainy video, but usually the girls were reticent to talk about what was happening. Only when she had adequately gained the girls' trust did they allow her to spend a night in their dormitory to see 'the sickness' for herself.

There were two dormitories, one for boys and another for girls. They were strikingly different environments. In the boys' room, the beds were arranged in orderly lines and were each neatly made up. The girls' room, on the other hand, was chaotic. Beds were unmade and scattered haphazardly around the room, the windows had broken louvres and the wardrobes appeared to be placed to create a barrier between the beds and the windows. The reason for these differences became obvious during the first night that Courtney spent at the school.

It happened early, before she had even gone to bed. The alarm went out that someone had collapsed and Courtney rushed to the dormitory, where she found a girl writhing on the ground, pinned down by her classmates. What Courtney witnessed over the course of the next few nights was unprecedented in her experience, but, when she described it to me, it was very familiar.

The girls didn't just have seizures, they ran around the dormitory in a frenzy and tried to push themselves through the broken louvres of the window to escape to the mountain. Fearing for their safety, their classmates, boys and girls, dragged them back and held them down on mattresses which had been taken off the beds and placed in a circle on the floor. Several people were needed to hold one girl down because, in the midst of 'the sickness', an afflicted person would develop an unnatural strength. Girls were restrained for their own protection, as the villagers feared they would run up the mountain and jump from the cliff if they were allowed out of the school. Some of the boys restrained the girls gently and carefully, while others giggled and used more force than seemed necessary. Once pinned down, the girls developed convulsions in which their limbs thrashed violently and their backs and necks arched unnaturally. Not everybody ran – some collapsed from the outset. The attacks lasted a long time, typically twenty minutes, although they could stop and start over several hours. Once they had recovered, the girls didn't remember much. When Courtney asked them what it had felt like, they told her that they had seen Granny, that she had come for them. They described her as a little old lady with white hair, dressed in white.

Sand Creek was not the first school to be affected by 'the sickness'. The phenomenon had been reported sporadically in Guyana over the previous ten years. Unlike grisi siknis, it was not endemic in the country and therefore did not come with any prescribed ritual or known cure. Western medicine had been the first port of call, but it hadn't helped. In one of the earlier outbreaks, before Sand Creek became involved, the authorities had even called in a US psychologist, Kathleen Siepel, to meet the girls. She diagnosed mass hysteria. The Guyanese people were very offended by this. The *Guyana Chronicle* caught wind of the psychologist's diagnosis and published an article headlined: *Statement by psychologist on mystery illness stirs controversy*. The Guyanese Minister of Education, Dr Desrey Fox, was quoted in the article as saying the explanation given by the US psychologist was 'typical of Westerners'. She declared the diagnosis derogatory. The local pastor also objected to the psychologist's assessment, saying, 'What that statement simply says is that the people's beliefs, that have been with them for generations, have no merit when compared to the modern-day ones.' The psychologist's declaration of facts, as she saw them, had taken a sledgehammer to generations of illness beliefs, tradition and spirituality within the community. The villagers, in turn, closed ranks and reinforced the scaffolding they had erected against her view. To their mind, the cure for the sickness involved removing the girls from the influence of Granny, which meant sending them back to their villages, where they duly recovered.

Mass illness outbreaks in schools aren't oddities or things of the past. They are not unique to any particular type of society, nor do they require a history of violence, poverty or hardship to occur. They can happen anywhere, to any sort of community, and do so on a regular basis.

Experts divide mass hysteria into two types. First, there is mass anxiety hysteria, which occurs out of the blue, without any preceding stressors needing to be present. It usually affects young people and happens in contained environments, like schools. It spreads through direct contact and comes and goes in a flash. A typical example happened on a November day in 2015, in Ripon, North Yorkshire. A student at a Remembrance Day service collapsed in an overheated assembly hall and forty other children quickly followed in his wake. By the next day, the children were better. Another example happened in Malaysia, in August 2019, when a schoolgirl started screaming and, almost immediately, screams rippled through the class and onwards, to neighbouring classes. It was over within hours. The El Carmen outbreak would probably have fitted into this classification had external forces not become involved to draw it out.

The second type is mass motor hysteria, which can affect people of any age, is more insidious in onset and lasts much longer.

Unlike mass anxiety hysteria, it typically occurs where there is a background of chronic tensions within a close-knit community. For example, in a struggling small town in the middle of the vast Kazakhstani steppe, in a time of political uncertainty and lack of control. Or in a US embassy under pressure, among people who have a reason to be fearful.

The big problem with mass hysteria, in any of its forms, is how it is perceived and understood publicly. There is a disconnect between the way mass psychogenic illness is defined and discussed by the small number of experts who study it and how it is understood outside those circles. In the medical field, it is regarded as a disorder that arises from group interaction and, as such, is sometimes, perhaps more appropriately, referred to as mass sociogenic illness. That makes it a social phenomenon rather than a truly psychiatric disorder. Unfortunately, in less expert hands, it is largely presented as a psychological problem, with all the focus on the individuals affected, and the essential role played by the community is almost entirely ignored. In its public persona, mass hysteria has become so attached to old-fashioned stereotypes, one-dimensional hypotheses of psychological trauma and clichés about young girls that portrayals of it have almost entered the arena of parody. In one writer's fictionalized account of the Le Roy outbreak, the girls' illness was ultimately diminished by centring it around a rivalrous argument over a boy. In the real world, a newspaper headline referred to the young women as witches. The teenagers in El Carmen were told they were sexually frustrated.

Aside from the fact that the clichés associated with mass hysteria are very demeaning to young women, they have the added effect of making the diagnosis untenable for social groups that do not fit easily into them. In both Cuba and Krasnogorsk, where the victims were older and some were male, the doctors rejected an MPI diagnosis outright and strongly supported the people's own formulation of what was causing their illness. They simply could not reconcile an MPI diagnosis with the type of people involved in those outbreaks because they were nothing like 'hysterical' young women. There is an irony in the way the people involved in the Cuba and Krasnogorsk events were portrayed, as opposed to the people in the school outbreaks: their stories actually fitted best with mass motor hysteria. Since that form of the condition is more likely to occur in individuals who are under psychological strain – as opposed to mass anxiety hysteria, which does not – one might argue that it was the two older, mixed-gender groups who would more accurately have been referred to as 'stressed'. And yet, the doctors and authority figures involved in those cases refused to acknowledge such a thing. It was only the schoolgirls – not the older people, not the men – who found themselves minutely examined for signs of being psychologically disturbed and who were repeatedly referred to as 'stressed'. If it was necessary at all, the opposite would have been more appropriate.

Aside from the misogynistic flavour that taints most stories about mass hysteria, I would also suggest that discussions feature far too many negative historical associations. It is irrevocably linked to witch trials, with mention of Arthur Miller's play *The Crucible* and its fictionalized account of the Salem witch trials never far behind. Many outbreaks also find themselves associated with accounts of some of the strangest illness clusters in history – among them, the dancing plague of 1518, in which hundreds of residents of Strasburg were driven to dance, with an estimated mortality rate of fifteen per day from cardiac complaints and sheer exhaustion. There were also the jumping Frenchmen of Maine, a group of nineteenth-century

Canadian lumberjacks who exhibited an exaggerated startle response, causing them to jump, and the laughter epidemic in Tanganyika, in 1962, which affected as many as a thousand people. These sorts of bizarre events are endlessly fascinating, but they only serve to create problems for those afflicted with the modern-day version of the disorder. To a US diplomat in Cuba, a Russian-deployed sonic weapon must have seemed relatively feasible and attractive in comparison to any association with these historical oddities. Other medical problems have moved on from old-fashioned associations. Tuberculosis is no longer a romantic disease afflicting poets. It can be talked about without recalling life in sanatoriums and without referring to it as consumption. Why hasn't mass hysteria moved on? I would argue that it is because people still struggle to appreciate the magnitude and reality of the interactions between mind, body and environment, so these conditions remain a spectacle at which to ogle.

It will already be clear how weary I feel of the constant name changes in the field of psychosomatic illness. It is as if people believe they can abolish judgement and stigma, provided they can find a sufficiently bland label. That said, mass hysteria remains a very problematic term. It is used in so many different ways that it has become not only confusing, but meaningless. In different settings, it could refer to social panic in the face of a socioeconomic crisis; excited teenagers at a pop concert; emotional outbursts in frightening situations; rioting; stampedes; panic buying; or mass shootings. The conflation of the medical disorder with frenzied, panic-stricken, emotional behaviour makes the medical diagnosis even harder to relate to for those affected – because, while they are aware of anxiety, it is the physical symptoms that dominate their experience. The El Carmen de Bolívar girls lost consciousness and had seizures. The people of Krasnogorsk fell asleep. Neither group felt especially emotionally overwrought.

It is also important to note that mass anxiety hysteria doesn't only happen in schools and doesn't only affect teenagers. It's likely that schools are particularly vulnerable to this disorder because they group people very closely, restricting their rights and limiting their autonomy. Young people's brains are still developing and the adolescent years are those in which peer pressure is felt most keenly, making schoolchildren at higher risk of social contagion. In the past, convents were also common sites for mass hysteria for the same sorts of reasons. Young women were isolated and subjected to extremely restricted lives, which is the atmosphere which fosters MPI.

Other environments conducive to mass hysteria are those in which there is reason to worry about environmental poisons or attacks – factories, for example, especially those using potentially dangerous chemicals. In June 2018, thirty workers fell ill with vomiting and shortness of breath in an e-cigarette factory in Salem, Massachusetts. Fire crews cleared the building and tested the air for chemicals. Having found no cause, they allowed the workers to return, only to be called back when more people fell ill. Atmospheric testing came up clear for a second time and the local fire chief ultimately attributed the outbreak to panic created by the smell of new carpeting.

The environment in aeroplanes and underground trains can trigger similar anxieties. Also in 2018, 106 people on a flight from Dubai to New York developed coughs and flu-like symptoms. While a virus could spread easily between passengers on a fourteen-hour flight, you would not expect people to become symptomatic within such a short time period. Later testing found no toxins or infections to explain the symptoms. Several people were taken to hospital in New York on arrival, but were ultimately given the all clear. It's not hard to see why concern about close confines and stuffy atmospheres can lead to events like this.

The outbreaks in Le Roy and Sand Creek have both been labelled as mass hysteria or mass psychogenic illness. Each of these health crises went on too long to be faithful to that classification. In a sense, they were mass hysteria events that evolved into functional disorders, but not necessarily *psychogenic* functional disorders. These are sociogenic phenomena. The explanation for each event lies in the society in which the outbreak occurred, not inside the girls' heads.

While the response to the El Carmen outbreak was shaped by isolation and a history of mistrust, the North American equivalent was sent into overdrive by the power of celebrity culture, sensationalist mass media and social media, all of which did the women a major disservice.

It is possible the outbreak in Le Roy would never have got out of hand had the first public meeting held at the school by the Department of Health provided more reassuring answers. The announcement that they could not release information because of privacy laws created a sense that something was being withheld. This led to scepticism about the conversion-disorder diagnosis, which in turn encouraged speculation about the cause, first locally and then in the public arena. The media showed no willingness to protect the girls from exposure. With mass hysteria providing great fodder for headlines, there was an inevitable snowball effect and, with so much media reporting being of a hasty nature, the quality of that reporting was frequently both weak and sensationalist. The need to be both brief and first left little time for deeper consideration of the subject.

And then came that growing scourge of the Western world: the inability to reliably distinguish opinion from fact. The involvement of unqualified media commentators was very problematic for Le Roy, because their speculation was given the same weight as expert opinion – or more, even, with Erin Brockovich's public popularity drawing a huge amount of attention. While I bow to Brockovich's achievement in Hinkley, we must bear in mind that she has no medical qualifications. The mere fact of her celebrity allowed her opinion to overshadow the doctors' informed diagnosis and advice. Meanwhile, other television presenters stirred the pot with conspiracy theories and ill-informed statements about what mass hysteria and conversion disorder meant to them, rather than how it is understood by experts. The effect of all this was to back the families into a corner.

The Le Roy girls were treated to every judgement available. Like the El Carmen families, they were put under a harsher spotlight than the US diplomats in Cuba. They were portrayed as being from poor, unstable backgrounds. The media pointed out the single-parent families and the relatively high unemployment of the town. One girl was reported to have a very strained relationship with her father, significant enough to be associated with some physical abuse. Much was made of the fact that another girl's mother had undergone several brain surgeries, when in fact she'd had some fairly low-risk operations for a non-life-threatening condition that was made to sound much more dramatic than it was. Every girl was cited as having a source of stress, such as a break-up with a boyfriend or an argument with a friend. Many accounts featured the loss of the Jell-O factory and the reduction in the town's previous prosperity, even though most of that had happened decades earlier. Publicly painting them in this disparaging way forced the girls and their families into a position of having to defend themselves; they emphasized that they did not experience their lives as being in any way bleak and therefore refused the diagnosis of conversion disorder. If having conversion disorder meant they were stressed, then it couldn't be right.

It is likely that, at the beginning of the outbreak, one of the girls began to genuinely suffer from a disorder like Tourette's, and that the social contagion to which teenagers are vulnerable created the spread of tics within a friendship group. Responsibility for the subsequent evolution of the crisis, however, lies firmly at the feet of the media. In fact, it was withdrawing the girls from the media spotlight that finally saved them and the town. The authority figures in the USA ultimately proved to hold more sway than those in Colombia could hope to. The neurologists looking after the girls took a firm stance and avoided being pressurized into pursuing an alternative diagnosis that they knew didn't exist. They condemned the media intervention and did their best to dispel the toxin theory and support the diagnosis of conversion disorder. The girls and their families were strongly encouraged to step back from the publicity and to eschew the involvement of outsiders. As soon as they did, the symptoms began to disappear and the residents of Le Roy, unlike El Carmen, could consign this story to their past.

Society in Sand Creek could not be further removed from that of an industrialized North American town like Le Roy. The customs, lifestyle, belief systems and family structures are radically different, and that is why Kathleen Siepel's distillation of the 'the sickness' into mass hysteria so utterly failed to work for the community. The girls of Le Roy and Sand Creek had almost nothing in common but the illness, so attempts to psychologize their suffering in a similar reductionist way was bound to fail. What had created 'the sickness' was integral to the community, not innate to the girls.

Mass hysteria, functional and conversion disorders are themselves culture-bound syndromes, but ones of the West. They have no meaning in Guyanese culture. 'The sickness' in Sand Creek probably arose out of deep spiritual beliefs and social factors specific to the region, with the triggering event being the disruption created by the change in the educational system. When the outbreak began, the village created a narrative to explain it that fitted their belief system. Living in Sand Creek, Courtney gained an appreciation for the community's way of life and learned to see their story in an informed way. Mass hysteria was a diagnosis that could only have come from someone who didn't speak the social language of the Wapishana.

'The sickness' could not be unravelled without understanding the traditional ways of the region. Kinship structure, relationships, learning and spiritual beliefs for the Wapishana differ in vital ways from those of the West. To the Wapishana, family is created by proximity. Sharing space and food brings people together, and living with somebody makes them kin. Courtney became a daughter to the family with whom she lived, just by being in their home and taking part in family life. Since actual physical closeness is so integral to kinship, it follows that prolonged separation from biological family threatens kinship. Western cultures place value on independence, but the Wapishana are seen through the prism of their social relationships. It is through someone's personal relationships that they come to know themselves, and that they are known by others. For this reason, people very actively circumvent interpersonal conflict to protect their relationships.

The complementary roles played by men and women in

traditional Wapishana communities were also disrupted by the new educational system. Women's lives are largely spent within the village, caring for the children, cooking, tending the garden and growing vegetables. Men's lives take them outside the village; they leave their home and go far away to get work, and go to other towns to find a wife.

Learning for the Wapishana also takes a different form to the structured, didactic learning of Western societies. Education happens through embodied learning; in the same way that emotion and ideas about illness are embodied, so too is knowledge. Embodied knowledge is gained through the senses. In this way of learning, experience is more important than instruction. That means that learning, like kinship, depends on proximity and immediate social interaction. Courtney came to understand the difference in learning styles in the kitchen of her Sand Creek home. Cooking proved to be a very important part of her integration into the community. In Western cultures, learning a new recipe often comes with clear instructions. When Courtney cooked local dishes for her host family, they made it obvious to her whenever she hadn't got the dish quite right, but instead of being told what exactly she had done wrong or being offered specific corrections, she had to find out for herself how to fix the mistake. She was expected to learn as they learned, by being in the kitchen and sharing the experience of cooking.

The Wapishana refer to learning physical skills as 'hand knowledge' and spiritual learning is known as 'eye knowledge'. What the school offers is referred to as 'brain knowledge'. Brain knowledge comes through the intellect. It is taught through order, timetables, curriculums and tuition. It has no need for community. Pupils sit alone at desks in an environment that promotes individual work. One can garner brain knowledge alone. Guyana values modernity and education every bit as much as any other country, as do the Amerindian Wapishana people. The government initiative to improve access to secondary schools, and the indigenous communities' enthusiasm for them, was a testament to that. The parents had been pleased that all their children would have an equal opportunity for schooling. This was advancement, but it also left them torn between two ways of life, with the more traditional way losing out. When 'the sickness' began, the school called a conventional medical doctor and a US psychologist before they did anything else. But the Amerindian way of life, their beliefs and spirituality, have never been represented in Western medicine or psychology, so, not surprisingly, both failed them. The only thing that made the girls better was to leave school, to leave formal education behind and to return home to live with their relatives.

The initiative to set up boarding schools and improve access to education was clearly an admirable one. It was welcomed by all parties, but it had unforeseen societal consequences. In a community where kinship depends so much on physical closeness, the girls' prolonged separation from their families had a huge impact on them, much greater than would have been felt in individualistic societies, such as in the UK or USA, where we expect our children to leave home, even encourage it. The impact on the girls was greater than on the boys because the role of Wapishana women lies at home in the village. Men had always left and interacted with outside communities, but women typically did not. What's more, the education the girls received in school did not necessarily reflect their needs. Separated from their female relatives, they missed out on gleaning new skills through embodied learning. Most would go back to the village when they finished school, but they wouldn't know how to cook the traditional recipes or care for the gardens and animals. Very few would progress to tertiary education, meaning that brain knowledge would serve little purpose for them.

The girls were torn between new and old ways of life, and the new way saw them lose as much as they gained. Sand Creek's sickness arose from a very complex embodied narrative in which the region's social structure and spiritual beliefs were central. Distressed by the loss of relationships with female relatives, they played out the story of the illness that had been coded by expectation in the neural networks of their brains. It was a medical disorder that sat well within their belief systems and it solved a problem. If leaving home was the cause, then going home was the cure. The girls had lost the crucial physical closeness of older female relatives and it occurred to Courtney that it was not by chance that the spirit who came to drive them back to their villages took the form of a grandmother.

I never had the good fortune to visit Sand Creek, but, in 2019, when the outbreak was long over, I did spend a beautiful autumn weekend in Le Roy. Visiting the place really gave me a sense of the disservice the media portrayal had done both to the girls and the town.

I took the scenic route to get there, driving through the quaint towns of the Catskills and the golden forests of the Adirondacks. I assumed that Le Roy would be a disappointment after the mountains' autumnal display, but it wasn't. After all I had read about the place, I expected it to be nothing more than a dying town full of boarded-up houses. There were some of those. The imposing building that had once been the Le Roy National Bank was long closed; its high, arched windows held signs advertising office space. Thrift stores hinted at a downturn in fortunes. But that was all offset by the grandeur of the wide main street, with its huge, federalist-style mansions set among mature green and golden trees. A creek cut the town in half, but that, too, was grander than expected. In my Irish vernacular, a creek is nothing more than a small stream, but this was a wide, fast-flowing river with manicured banks, one of which was home to a mini-version of the Statue of Liberty. The creek had featured in some of the articles I'd read about the town. Children swam in it in the summer, and contaminated water was one of the many causes suggested to explain Le Roy's mystery illness.

Besides the natural beauty of the place, there was also a rich history to explore. Next to the creek is the Woodward Memorial Library, all pillared grandeur, set off by a hundred-year-old English beech tree. In front of it I found a sign which read, *Ingham University Campus, Site of First Women's University.* Le Roy, it turned out, was the location of the first chartered women's university in New York State. Once upon a time, at least, Le Roy was a feminist town.

Eating chicken pot pie at a table with a chequered cloth in a converted railway-station restaurant, I was filled with the warm personality of the place. Thinking back to the news coverage of the outbreak, it felt almost as if the convenient negatives from the history of Le Roy and the lives of the girls had been cherry-picked to suit the depressing, hysterical narrative, so that all the positives of the place and the people had been lost.

The Le Roy girls had insisted they were not stressed or depressed – no more than the average person, at least – but it just didn't suit the public or media version of the disorder to accept their view of their own lives. That has been the experience of many people with functional disorders, too – they are forced to insist that stress is not at the heart of their problem. I would suggest that those so determined to refer to the Le Roy girls as being in denial about the cause of their symptoms were actually pointing the finger of blame in the wrong direction. The extent of this outbreak had much more to do with the social response to the problem – namely, the finger pointers themselves – than it did to do with the psychology of the individuals.

There is definitely something to be learned from what happened in Le Roy, but I fear that those who need the lesson most have already moved on and will not benefit from hindsight. Those who would not accept the reality of functional and related disorders will almost certainly still refuse to accept it. Six months after their investigation started, Brockovich's team held a meeting in the town's American Legion to announce their findings. There was an agreement that the train-crash site had been badly managed and that there had been a chemical spill, but there was no evidence that it was a threat to the town. There was no toxic plume hanging over Le Roy. Brockovich did not attend the debrief meeting in the Legion, citing ill health. In other words, she personally did not undo the panic that her involvement had triggered – that was left to one of her investigators, who confirmed that the creek and underground water system flowed directly away from the train-crash site, in the opposite direction to where the school lay. It makes one wonder: if someone had looked at the map more carefully before Brockovich's dramatic broadcast with Pinsky, might the conspiracy theory have been knocked on the head, cutting short the girls' suffering? Their doctors' diagnosis had been dismissed as too quick and too convenient, but I would love to know if Pinsky and Brockovich ever realized what their own haste and fevered excitement had done. Brockovich never revisited the site and never admitted she had been wrong. In fact, she did almost the opposite. After drawing a blank, she did not back down, but instead threatened to continue her investigation, although ultimately she did not.

When the outbreak in Le Roy exploded into the media, it did so loudly, but the fizzling out of the toxic-spill theory was much lower key. No conspiracies were uncovered, there were no cover-ups – and this, let's face it, is not the stuff of headlines. The nature of headline-grabbing journalism is that there is no requirement to fix these types of mistakes. By the time the truth has been uncovered, people have usually already moved on to the next big story. Once all the fuss and multiple investigations were over, conversion disorder remained the most plausible explanation for what had happened in Le Roy, and no other realistic alternatives emerged. Most people who read the original newspaper articles in the USA and around the world will probably never have heard how the Le Roy story ended, for the simple reason that, by then, the story had stopped being newsworthy.

During my visit to Le Roy, I spent time with a local journalist, Howard Owens, who had worked on the story and who had been at many of the public meetings with the families and other members of the media. He recalled that, when the conversion-disorder diagnosis was first mooted, he had never heard of it before, so he had done a quick Internet search. 'I was sitting at a public meeting and, when it was mentioned, I just had a quick look on my phone. That search suggested it was nonsense,' he said. 'I didn't believe it was a thing.' A large proportion of the media seemed to agree, which is why the diagnosis was presented as if it was a dismissal. Clearly, the medical community needs to work harder. We cannot expect the untrained public to understand these disorders if we doctors have not yet caught up.

Howard had been present when Brockovich's team, along with several international news teams, had had a heated confrontation with school superintendent Kim Cox on the edge of school grounds. I asked him how he felt in the aftermath. As a local, intimate with the community, he had seen the story play out right to the end.

'If I knew then what I know now, I would have done things very differently,' he told me.

Howard had disbelieved the conversion-disorder diagnosis because, like most people, he was not familiar with it and it sounded like a non-diagnosis to him. He had also seen the discarded barrels at the train-crash site and they seemed to be a compelling reason to be genuinely concerned about an environmental poison. Nobody would want that on their doorstep. But, as the story unfolded, the toxin theories stopped making sense, and the more Howard found out about functional disorders, the more it seemed a reasonable diagnosis. Few other journalists returned to Le Roy to get the ultimate answer and learn from it.

I had a strong sense of what I thought should have happened differently, and I suggested it to him: 'Maybe Erin Brockovich should have held off speaking on television until she had visited the site. The investigation, at least in part, should have come first.'

Only one television crew that I know of went back to Le Roy for a follow-up – a Japanese news team, who pride themselves on exposing the truth and make a point of clearing up any misinformation for their viewers. Pinsky also ultimately gave some lukewarm support to the diagnosis of conversion disorder, although he still claimed that one could never be certain. When asked if the media attention benefitted or harmed the community, he acknowledged some detriment, but made the point that, without the media, the town would have missed out on crucial interventions – chiefly, that of Erin Brockovich.

By Western medical classification, Krasnogorsk's sleeping sickness, El Carmen's seizures, Le Roy's tics, Cuba's Havana syndrome and Sand Creek's sickness all amount to more or less the same medical problem: mass hysteria. A complex, tangled web of social, environmental, medical and psychological factors created the specific symptom trajectory for each group, but this web was reduced by many to contagious panic, fear, anxiety and psychological fragility. As such, the diagnosis was rejected. The characteristics that made each outbreak a distinct entity were lost by this classification. The greatest difference between these mass events was not in the people affected, it was in the society in which they lived. It is the social differences that are key to understanding the cause of and suggesting a solution to mass sociogenic illness.

Mass hysteria is a magnified version of all that is wrong in the way we perceive and discuss psychosomatic and functional disorders. Stereotypically, the condition is rejected as a diagnosis for men and caricatured for young women. There is a great deal of work to do to improve our understanding of how these disorders develop, but there is one initial step that should not be difficult – surely, it is time we stopped resurrecting the centuries-old tropes of witch trials and Freudian hysteria every time a mass outbreak occurs or a young woman faints.

8

Normal Behaviour

Normal: Conforming to the standard or the common type; usual.

In Korea, there is an illness called hwa-byung, meaning 'fire illness'. It is one of the conditions referred to as a culture-bound syndrome or folk illness. The main symptom is a sense of heat or burning all over the body, accompanied by a variety of other somatic complaints, like chest pain and shortness of breath. In a Western medical setting, a person with this constellation of symptoms might be met with reassurance, but would also have a high likelihood of being offered a range of blood tests. They might also be referred for nerve conduction studies, used to assess the integrity of nerves, since neuropathy can cause a burning sensation in the skin. Or, if the chest symptoms are prominent, cardiac investigations could be recommended. Alternatively, if the problem was assessed to be psychosomatic from the outset, the affected person might find themselves being referred to a psychiatrist.

Hwa-byung has cultural meaning to Korean people that a Western doctor would have difficulty appreciating. It affects middle-aged women in particular, and is associated with the stress induced by marital conflict and infidelity. Like grisi siknis, hwa-byung is a language of distress, understood by the community that speaks that language. The specific symptoms are not meant to be taken literally; they are a metaphor for a particular type of psychological suffering. Hwa-byung is an acceptable way of asking for support.

The *DSM-V*, the psychiatry bible that catalogues mental disorders, only specifically names culture-bound syndromes belonging to communities that do not have English as their first language – susto, shenjing shuairuo, khyal cap, nervios, dhat. Cultural concepts of distress are defined as the ways that cultural groups experience, understand and communicate suffering, behavioural problems or troubling thoughts and emotions. So, if the *DSM-V* does not specifically name any culturally defined disorders that originate in English-speaking, industrialized, Western communities, does that mean we don't have culturally shaped illness? Are we so open about our suffering that we have no need of metaphors?

Some Western cultures do have medical complaints that are unique to them. In France, there is a condition referred to as *les jambes lourdes* – meaning 'heavy legs' – that is not commonly seen in other countries. The medical literature for this condition is almost exclusively in French. It is attributed to venous insufficiency, which is believed to lead to fluid pooling in the legs, resulting in heaviness and swelling. Apparently, if a person goes to a pharmacy in France and reports having heavy legs, they will be directed to a shelf loaded with products said to alleviate the symptoms. One commercial website selling a variety of creams and gels for *les jambes lourdes* says it affects up to one in three women. And yet, 'heavy legs' does not exist as a disease category in the UK.

Les jambes lourdes would not be referred to as a folk illness or culture-bound syndrome in France, because these are terms more often used to label people outside of one's own cultural community. It is very difficult to either spot or talk openly about the cultural idioms within one's own society, partially because they are not recognized as such, but also because they are presented as biomechanical illnesses and to say otherwise risks forcing something that is being hidden for a purpose into the open. I am a doctor trained in the Western medical tradition, Irish born and living in London. These are the main cultural factors that influence my own health and illness beliefs, and I am indoctrinated to use that cultural language when talking about illness. Like many Western doctors, I medicalize feelings and behaviour. People come to me so that I will do that for them – give them a medical explanation for their suffering – but, in truth, I worry all the time that what I'm doing, faithful as it is to my training and welcome as it may be to my patients, is wrong and potentially harmful.

Sienna came to my clinic for a second opinion. Although she was only twenty years old, she had a long list of medical diagnoses. According to her referral letter, she had joint problems, headaches, memory disturbance, low blood pressure, a tendency to faint and a sleep disorder. None of these were the reason she had been referred to me – that was because she was having blank spells and had become convinced that she suffered with epilepsy. Another doctor had told her this was not the case, but she and her family doubted that opinion.

'You've got to understand the sort of girl Sienna is,' her mother told me, almost before I had heard any of the story. 'She is very talented, a very high achiever, and these petit mal are ruining her life. We need them taken seriously.'

'Petit mal' is an out-of-date medical term used to refer to a very specific type of epileptic seizure, mostly seen in small children. One of my many flaws as a doctor is an impatience when people describe their symptoms to me using technical terms – I would rather hear a story told in plain descriptive English. I knew I was off to a bad start when I said as much to Sienna and saw her mother's startled expression.

'You don't have to convince me you're suffering, Sienna –' I directed my comment to her, rather than her mother – 'you wouldn't be here if you weren't. All I need you to do is describe your symptoms, from when they began until now, but please don't use medical terms, because that can be confusing.'

'But this is how we always explain it!' Sienna said, obviously affronted. She was seated with one parent on either side, and the three exchanged some none-too-impressed looks at my interruption.

'The details are in the referral letter, surely? We've given you all her previous doctors' letters,' her father said sternly.

'The letters are useful, thank you, but, since this is a second opinion, I want to wipe out what's happened before. So, if Sienna could just go back to the start of her story and use her own words to tell me what the problem feels like, that would be great.'

One medical study showed that, in the case of dissociative seizures, patients are more likely to describe the consequences of the attacks and the places where they occurred, and are less able to provide subjective details of the experience. People with epilepsy, on the other hand, are better able to focus on their seizure symptoms. In my experience, that applies to many functional disorders. Sometimes, when talking to people with functional disorders, it feels almost as if they want their suffering acknowledged more than they want treatment. A diagnostic label does that, but it has to be a good enough diagnosis. Epilepsy would certainly be serious enough to explain how bad Sienna felt, but her previous doctor had insisted that she didn't have it.

'Can't you just read the letters?' Sienna asked.

'I have read the letters, but now I need you to describe your symptoms in your words. Otherwise, I'm just copying the last doctor, and, let's face it, you didn't agree with him.'

Reluctantly, Sienna went back to where it all began, a year before. Slowly, despite her struggle, I began to get a sense of how her illness had unfolded. The problem started in university, when she noticed that she couldn't follow the thread of lectures. She also struggled to keep up with detailed conversations. As she told me about it, she peppered the story with interjections about how doctors had dismissed her or friends had abandoned her. I tried to be patient. Each time she meandered off the subject, I led her back to the point as carefully as I could, but I knew we were both feeling frustrated. Eventually, I grasped that she had first suspected a problem when she fell behind with her studies, but only later did she attribute this to blank spells, which she believed were affecting her concentration. At some point, she developed a certainty that the blank spells were due to epilepsy. They had been infrequent at first, but they increased in occurrence to such an extent that eventually they were happening multiple times a day, and were always most prominent in lectures. Chatting with friends, she could ask them to repeat things, but, since she couldn't interrupt or ask the lecturer to slow down, she felt herself getting lost. She had partially solved that problem by recording presentations, but group tutorials were very challenging. She missed so much of the discussion that she didn't feel able to ask questions in case she accidentally repeated something that had already been covered. As it evolved, these difficulties bled into every part of her life. She got into arguments with friends and family, who insisted they had told her something, while she insisted they had not.

'What does it feel like when you realize you've missed something?' I had to ask her this several times, in several different ways, before she found words to explain the feeling.

'Like I've just been switched off.'

'So, are you completely blank? Do you ever have any internal feelings that warn you it's about to happen? Are you even partially aware?'

'The best way to describe it is that it's like being caught in a dream. Or spaced out, as if I've had a drink and am getting a bit tipsy.'

She described it very eloquently, once she gave herself time, I thought.

'These days, is it still more likely to happen in certain situations?' I asked.

'If the lecturer talks too quickly, or if there's too much noise in the room. I have an extreme sensitivity to sound, at times. I can't function in a room where there's too much noise. I think my brain can only process one set of information at a time.'

'I can see when it's happening to her,' her mother added. 'Her whole expression changes. She's such a bright girl, but she looks empty.'

'Can you bring her out of it?'

'Sometimes. If you stimulate her, she sort of startles and wakes up.'

Our conversation had become less uncomfortable as we progressed through the story. While unhappy with my style of questioning at the start, the family seemed to warm to me as we went through the details. I asked about Sienna's other health problems. It seemed too much of a coincidence that a young woman would have multiple unconnected medical conditions.

I learned that, as a young teenager, she had complained of knee pain, which, after several visits to various doctors, was attributed to joint hypermobility. Later, after a series of dizzy spells, she learned she had low blood pressure. That had led to lots of medical tests and a diagnosis of postural orthostatic tachycardia syndrome (PoTS). This is a medical disorder that manifests as a fast heart rate provoked by standing or sitting up from a lying position. The main symptoms are dizziness, fainting and fatigue. She also had irritable bowel syndrome. She took homeopathic remedies for a sleep disorder. She described waking two or three times per night and never having an entirely unbroken night of sleep. She had seen numerous doctors in the previous five years.

I was interested to note that Sienna had made quite a lot of life concessions to accommodate her many medical diagnoses. She had all but given up sport because of her joint hypermobility and ate a very restrictive diet to alleviate her irritable bowel symptoms. Her parents had put blackout blinds and some soundproofing in her bedroom to help her work and sleep. She tried not to go out alone, in case she felt unwell, so friends or family were almost always with her. Because of her noise sensitivity, her school had sometimes allowed her to sit exams separately from everybody else, with only an invigilator present. As she described her life, I could not stop myself wondering how she would manage when she left formal education and the care of her parents, and entered a world that would not always bend to accommodate her as her current world did.

'I have read quite a lot about epilepsy and I know that petit mal causes daydreaming just like this,' Sienna's mother said. 'A woman in my office has a son with epilepsy and saw Sienna have one of her attacks, and she says it's just like what happens to her son,' she emphasized.

I had finished my questioning and had asked them to let me know if there was anything they thought I had not fully understood.

'We just want a coherent explanation and treatment,' her father added. 'She's due to start her second year of university. Her tutor says she's good enough to get a first, but that isn't going to happen while this is going on.'

'I'm sure it's epilepsy,' Sienna said, driving home the point. 'I know the last doctor said it wasn't, but I don't think he looked hard enough. He did hardly any tests.'

I did not think that Sienna was describing epilepsy. In fact, when she described her difficulty concentrating in noisy rooms, the panic she felt when she realized she had lost her place in a lecture, her occasional need to read the same page three times, I could not escape the thought that each of those things had happened to me many times. I kept that thought to myself; it seemed unlikely that normalizing Sienna's experience by comparing it to my own would be appreciated at that moment. Some people want to be reassured, but not Sienna.

I suggested I arrange a series of tests, and I saw the family exchange happy glances, because that's what they had wanted all along. 'But I do need you to know that I consider epilepsy a very unlikely diagnosis,' I said. 'The symptoms you described would be very unusual for that.'

'That's what our doctor said when we asked for Sienna to be tested for PoTS,' her mother told me. 'They said there was nothing wrong with her, that feeling dizzy all the time was okay. The first test for that was normal, too. We had to pay privately to get a second test, and that proved it. If we've learned anything in the last few years, it is that doctors are wrong all the time.'

'Sure,' I agreed. 'I may be wrong.' I knew that epilepsy would be fairly easy to rule in or out as a diagnosis, so we didn't need to be in agreement at the first consultation. Blank spells due to epilepsy are accompanied by some pretty dramatic abnormalities in the brainwaves, so I would let the tests decide.

I met Sienna next when she was admitted to the ward for monitoring in the video telemetry unit. For a full week, she wore EEG electrodes to monitor brain activity and ECG electrodes to measure her heart rate. Another electrode assessed muscle tone, while a pulse oximeter measured the levels of oxygen in her blood. The nurses did regular blood-pressure checks. Two video cameras meant that Sienna was watched at all times from different angles. Nurses were ready to go into her room whenever she pressed the alarm to report that she felt unwell, or if the nurses noticed something untoward. If she had a blank spell, part of the nurses' job was to ask a series of questions and give a range of instructions to assess Sienna's level of awareness and ability to interact. With all of that in place, we only had to wait. Sienna was asked to keep a diary of how she felt. This set of tests had to be definitive, so I was covering all the bases. At the end of the week, I would collate the information and correlate what we recorded with her experience of her symptoms.

We were only halfway through the week when the nurses alerted me to the fact that Sienna was having difficulty complying with the routine of the ward. She complained bitterly of broken sleep and refused breakfast because it was served too early in the morning. When the psychologist called to see her at 9 a.m. to perform detailed memory tests, Sienna sent them away, saying that she needed an afternoon appointment because her sleep disorder meant she was not a morning person.

'Look at this.' The EEG technician, whose job it was to review every second of the video and EEG tracing, drew my attention to a section of the recording. Sienna's mother was feeding her breakfast to her, as if she was a baby, offering her bites of a piece of toast as Sienna lay in bed.

'Why is she doing that?' I asked.

'She says she's too dizzy to sit up until the afternoon. Says her blood pressure is too low in the morning.'

'She told me she had PoTS, but never mentioned dizziness of that magnitude when I saw her in clinic,' I remarked.

'I don't think it's this bad at home; it's a problem here because the room is too hot.'

I decided to gather as much information as possible before trying to work out the reasons for this behaviour. By the end of the week, we had accumulated over 150 hours of waking and sleeping brain rhythms, and had a very extensive diary of symptoms to discuss. While reading, watching television and in conversation with her visitors, Sienna repeatedly reported blank spells and losing the train of conversation for several seconds at a time. Sometimes, when the nurses went to check on her, she didn't acknowledge them or couldn't answer questions for up to a minute. I looked at her diary, in which she had written, 'spaced out', 'blanks', 'can't think' or 'feel weird' dozens of times every day.

I looked at the many sections of video and EEG during which Sienna had reported feeling unwell, and found that, even when she was uncommunicative, all her physiological measures, in particular her brainwaves, showed a normal waking pattern. A few times, Sienna's mother pressed the alarm to say that she was certain her daughter was having a blank spell. Looking at the video of those events, Sienna appeared perfectly well and it was hard to see what her mother had detected in her.

'Polysomnography was normal,' the EEG technician said, letting me know the result of the sleep analysis. Sienna had not asked me to comment on her sleep disorder, but, with all the equipment in place to test sleep, it seemed silly to neglect that detail. There was nothing odd to find. While Sienna did rouse from sleep a few times a night, as she said she did, the arousals were always very brief; she had progressed through normal sleep stages and had slept for a normal amount of time overall. I looked at her blood pressure and heart rate readings too, mindful of her unwillingness to sit upright until lunchtime. Throughout her stay, her blood pressure was at the lower end of the normal range, but still normal.

I had let Sienna's family know that I would meet them on Friday morning to discuss the results. Since my ward round was at 10 a.m., I found Sienna lying down in bed and reluctant to sit up for our meeting. Her mother sat by her and the two held hands. Her father stood, leaning against the wall, arms crossed and stonily silent.

'Does this happen a lot?' I asked, referring to the dizziness that made her too scared to sit up to speak to me.

'Not as bad as this.' Sienna smiled. 'This room is too warm. The window doesn't even open.'

She had a point. Like many hospital rooms, the heating hadn't adjusted to some unseasonably warm winter weather, and the fourth-floor windows were sealed shut.

'Sorry about that. It is stuffy in here.'

'I know why I'm dizzy, so I don't mind. I have PoTS,' she said lightly. 'But what did the tests show? I had some really bad seizures this week. Did the nurses tell you?'

'Yes, they told me everything, and of course I have your diary, too,' I said, and then went on to explain how the test worked. I had looked for anomalies in her brainwaves and other readings whenever she reported feeling forgetful or losing time. In association with any type of loss of consciousness, there is always at least some change in the brainwave pattern, irrespective of the cause. But, no matter how unwell Sienna felt, even when she was completely unaware of her surroundings, her brainwaves showed a normal waking pattern. In fact, all the parameters I had measured were normal.

'They can't be,' Sienna said, with utter disbelief. 'I have epilepsy, I'm sure I have. Did you watch me during the memory tests? The psychologists gave me an easy story to remember and I couldn't remember any of it.'

In fact, the psychologist agreed that Sienna had struggled with the tests, but thought it was because Sienna had been so anxious that it had affected her concentration. Sienna had difficulty taking in information, rather than retaining it. She was simply too flustered to perform well.

I explained that the normal tests meant she did not have epilepsy.

'Maybe she doesn't have epilepsy, but surely it could still be something else,' her mother said.

'Clearly it is something else – it's a problem, whatever it is – but the normality of the tests fits best with these blank spells being due to a psychological process called dissociation, rather than being an irreversible brain disorder. These are episodes we refer to as dissociative seizures.'

I described dissociation and explained it as something to which we are all prone, especially when overloaded or distracted by life's problems. It's the moment you can't take in anything your partner is saying because your head is too full of other thoughts. It's forgetfulness on a bad day. It's dizziness in a stressful situation. It's the time you couldn't concentrate enough to follow a simple instruction. I explained that normal psychological processes like dissociation serve a vital purpose in life: they stop us from becoming overwhelmed and, in some people, protect against painful thoughts. However, dissociation, like every other bodily function, can go wrong, and, when it does, it can stop being normal and become a problem.

'I don't know,' Sienna said, and she exchanged an uncertain glance with her mother.

This is the moment I dread in consultations, when I have taken away the disease the person had expected and put a psychological mechanism in its place. For many, that is seen as a poor substitute.

'This is treatable,' I reminded her. 'While very troubling now, there are ways to overcome this experience.'

Then, help came from an unlikely place. Sienna's father, who had been standing quietly in the background, suddenly interjected, 'I think she's right. I've had that too, Sienna, and you did once tell me that your head felt too full to take in new information.'

'Yes, it does feel like that sometimes,' she agreed.

'And, if this is treatable, then we should at least do what the doctor is suggesting.'

For the whole conversation, Sienna had remained lying flat and I had felt I was looming over her. Her father's support of the dissociation diagnosis buoyed me up and I wondered if this was an opportunity to address her other medical problems, some of which had also been contradicted by the test results. She had only come to me to discuss the possibility of epilepsy, but if she could accept the blank spells were a manifestation of dissociation, perhaps she could see that at least some of her other medical complaints might have a similar cause. Dizziness is a very common symptom of dissociation, and her experience of poor sleep had been objectively contradicted by the tests.

'People's perception of how much they sleep is not always accurate, and Sienna may not be as bad a sleeper as she thought,' I suggested.

'I wasn't looking for treatment for that, anyway,' Sienna said, dismissively.

'Sure, but I thought it might be good to learn that you don't appear to have a sleep problem. You've had seven nights of good quality sleep while you've been here.'

'That's good, I guess,' she said, still showing little sign of being interested in this news.

'It's the seizures we were worried about,' her mother interjected, supporting her daughter.

Not knowing when to stop pushing my luck, I was still interested to see if Sienna really needed to spend the whole morning lying down. She attributed this requirement to blood-pressure problems related to her PoTS diagnosis, but that would be an easy theory to test. The reason a person with PoTS might be unable to sit or stand was because their body was unable to sustain their blood pressure in the upright posture.

'While you're still hooked up to all this equipment, I wonder if this might not be a good time to get you sitting up and standing up? Maybe your body can handle that better than you think, too?'

Sienna looked very doubtful, but her father helped out again: 'You might as well; you're unlikely to ever be monitored in this much detail again.'

The nurse accompanying me fixed a blood-pressure monitor in place and I turned the screen that showed Sienna's EEG and heart tracings so I could see both clearly. With some reluctance, Sienna sat up slowly. Dizziness and collapse in PoTS are caused by an abnormal autonomic response to a change in posture, which causes the heart to race and ultimately the blood pressure to fall as a person sits or stands from the recumbent position. Sienna reported feeling dizzy, but succeeded in sitting fully upright with only a small increase in heart rate and no change in blood pressure. Relentless as ever, I asked her to stand.

'I'm afraid to stand,' she said, her voice full of fear.

'The moment you feel sick, I'll lie you down again,' I said. 'For the completeness of testing, I think it would really help to see the effect of a change in posture on your body.'

Painfully slowly, she swung her legs over the edge of the bed. Her mother stood, watching anxiously.

'I won't let her fall,' I said, trying to reassure them both.

'You just can't imagine what it's like to watch your daughter go through something like this.'

The nurse and I flanked Sienna, ready to support her if she couldn't stay upright.

'Oh no, my head is spinning,' she said, as she gradually came to a standing position.

'Do it slowly. If your blood pressure gets too low, your body will almost certainly detect it and fix it.'

She steadied herself. I could see Sienna's heart rate and brainwaves pulsing along normally on the screen in the corner of the room. That is not to say Sienna's caution in standing was unjustified. As soon as she was fully upright, her knees buckled. She closed her eyes and her body went limp. The nurse and I supported her to fall backwards, so she landed on the bed, then the nurse deftly lifted her legs to bring her to a comfortable position. As soon as she was lying down again, she woke and started to cry.

'I told you that would happen,' she said, between the tears.

'You did. Thank you for letting us do it anyway. It helped for me to see it for myself.'

Sienna's heart rate had risen slightly as she stood, but all other physiological parameters were normal. The fear, in PoTS, is one of fainting due to the brain being deprived of blood and oxygen as the blood pressure falls. The EEG (brainwaves) slow dramatically during a faint. They had not slowed when Sienna flopped back onto the bed, so this was not a faint. This dramatic response to standing was related to an expectation, rather than a pathological response in her body. Ultimately, this too was due to the brain shutting down through dissociation, triggered by fear. It was also something that was escalating through habit. She had come to expect that she would feel dizzy in the morning, so she looked out for it. She associated dizziness with low blood pressure and believed that standing when she felt dizzy would inevitably lead to collapse. So, when she tried to stand, her expectations overwhelmed her nervous system and fulfilled her prophesy.

It was becoming clear that many of Sienna's medical complaints had a functional (psychosomatic), rather than pathological, cause – the tests had supported a functional diagnosis for her blank spells and for her inability to stand in the morning. The normal polysomnography, which contradicted her experience of her own sleep quality, suggested that hypervigilance to small arousals had led her to develop undue concern about her sleep. It made sense that these problems were related. They gave a single unifying diagnosis, rather than lots of unconnected diseases. I feared that, if her tendency to develop functional neurological problems and accumulate disease labels was not adequately addressed, it could be progressive. I have seen many people with a story that started out like Sienna's ultimately develop serious long-term disability.

Once Sienna started to feel better, I discussed my concern that PoTS could not account for the degree of dizziness she described.

'I always faint if I try to stand up when my blood pressure is low in the morning,' Sienna contradicted me.

'Except that wasn't a faint, Sienna. To faint, your blood pressure would have to fall sufficiently to deprive your brain of blood and oxygen. But I could see all those measures as you stood up, and they were normal.'

'But I have PoTS,' she said.

'Perhaps, but what happened just now was not due to PoTS.'

What ensued was a long conversation about being too observant to small bodily changes, and advice on how to stop that. I suggested she slowly trained her body to become used to standing in the morning, whatever the weather or ambient temperature. She listened in a distracted way, occasionally making unconvincing sounds of agreement. I had the sense, once again, that she had come to see me for one reason and one reason only, and any attempt to intervene on anything else was overstepping the mark.

She agreed to see a psychologist, who I promised would try to help her break the cycle of her symptoms. I felt relieved that, by the end of the conversation, she was happy to accept the concept of dissociation as a cause for her blank spells and had relinquished the certainty that she suffered with epilepsy. I hoped the psychologists could teach her to react differently to bodily changes.

I didn't see Sienna for a while, after that. When we next met in clinic, she arrived with her parents and a smile on her face. She was feeling a lot better, she told me. She had read some more about dissociation and it made sense to her. She had talked to her tutors about the issue and they were giving her extra attention to help her catch up on work.

'Can you provide me with a letter for my university, please?' she asked. 'I need an extension on the deadline for my dissertation.'

'Certainly – you've had a rough year. What would you like me to say?' I asked.

'Can you tell them my diagnosis and explain I need a quiet environment and that I will need extensions for assignments?'

'I can say that you spent a lot of time in hospital recently, and that interrupted your work, but do you really think you need more time for future assignments? My hope is that the work you are doing with the psychologists will get rid of the blank spells, and I think your expectation should be that you'll get back to normal and be able to work in the same way as your classmates.'

She did not look happy at that.

'If you just give me a letter letting them know I've been sick this year, I can ask my PoTS doctor for a letter for next year.'

I remembered what I had thought when she first told me her story – that she would struggle when she entered a world that wouldn't bend to suit her needs.

'Is it possible that this course might be wrong for you?' I ventured, avoiding saying too explicitly that I thought she might be stretching herself beyond her abilities.

She was having none of that. 'I could do this course, no problem, if I didn't have so many medical issues.'

I looked to her parents to see how they felt, but they said nothing, so I backed away from the subject and quickly typed a letter asking her course leader to keep in mind her illness and the time she had lost through hospitalization.

As the family left, they thanked me warmly. I had done exactly what they wanted. I had tested Sienna thoroughly and, although I had given a different diagnosis from what they had anticipated, it was a firm diagnosis and they had left satisfied. Sienna was getting better, her family was happy – so why did I feel guilty? I knew why, of course. I had done everything that my job required of me, I had done what my colleagues would do, I had followed the conventions of my training, but the truth was that I wasn't entirely comfortable with all of those conventions. I knew I had withheld a much franker version of my opinion. Over thirty years of working in Western medicine, I had learned to comply with the increasing need to label everything as if it was an illness, but deep down I believed that this did many patients a disservice. If I had been more truthful with Sienna, I would have said I thought her symptoms indicated a difficulty coping with escalating academic pressures. I would have said they were not an illness, but a sign that the life she had chosen was impacting negatively on her. If she was struggling to achieve her goals, maybe they were the wrong goals. But, in Western society, when things are going badly for a person, medical explanations are often sought because they are found to be more palatable than psychosocial explanations. Western medicine has, in a sense, learned to comply with the needs of the people. Thus, the lines between behaviour and illness, normal and abnormal – even the demarcation between disease and health – have become so blurred that it is possible to give an illness category to almost any person. Once that is done, a person becomes a patient.

I had given Sienna a medical label to account for her struggles. I had followed that up with a letter that required her university to make life easier for her. Was that really what she needed? Twenty years ago, I would have explained the phenomenon of dissociation without using the actual word. I might even have told her, in a much less roundabout way, that I thought she'd taken on too much. These days, it's hard for a doctor to do that. When people come to specialist doctors like me, they have made it past the family physician, who will almost certainly have offered reassurance. They have been referred on because that strategy hasn't worked. The most satisfying doctor–patient encounters are those in which a diagnosis is offered and accepted. Insurance companies and sick notes and hospital coding systems all like doctors to put patients into a classification.

As she was a person presenting for a second opinion to a tertiary referral centre, I assumed I could not satisfy Sienna unless I gave her a definite diagnosis she could trust, so that's what I did. I could call it a success, but at what long-term cost? Sienna showed signs of embodying medical diagnoses, so would dissociation ultimately disable her, the way the PoTS diagnosis had? After she had been given that label, she had enacted it so completely, it had made her worse instead of better.

Dissociation is a normal phenomenon, in which it can be hard to know where normal stops and abnormal begins. That is the same for sleep – we all sleep differently, so, while creating norms within a population is practically possible, it is not as easy to determine exactly the right amount of sleep for each individual. I suppose one could say that abnormality begins the moment a deficiency in either sleep or dissociation leads to significant difficulty functioning in the world. If that is the case, the extreme dissociation Sienna experienced did represent illness and my course of action was the correct one. But that does not allow for the potential effect of labelling a person as 'ill'. I could have tried harder to avoid a label and frame the discussion in terms of Sienna's interaction with the world. I chose to offer Sienna a medical classification because that was the only outcome I thought she would accept. In doing so, I accepted the risk of Hacking's 'looping' and 'classification' effects and gave her an opportunity to embody a new sick role. That is Western medicine's culture-bound syndrome – we make sick people. We medicalize difference, even when no objective pathology is available to be found. Sometimes, we are right to do so, but we are also wrong more often than people realize. And while resignation syndrome and Havana syndrome affect small groups of people, when Western medicine introduces new classifications to the world, it does so on a grand scale, without people's consent, and there is something frightening in that.

The presence or absence of illness and disease are not immutable scientific facts, as many suppose them to be. Certainly, some diseases are blatant – you either have them or you don't. But many are not as black and white as that. At what value does pathologically high blood pressure start, and is it the same for everyone? Most biological measures do not have a single correct value; they have a range of values and everything within that range is normal. In deciding who is diseased and who is not, there is the problem of knowing where to set the cut-off point between normal and abnormal. How low or high do hormone levels have to reach before they can be classed as outside the acceptable normal range? How much dissociation is too much? Height, weight, heart rate, blood sugar, haemoglobin level and numerous other biological measures are given limits by expert committees, and those limits decide who is sick and who is healthy. With no absolute right or wrong answers, scientists use their experience and knowledge to place those limits, but an element of that is inevitably arbitrary and the process often favours over-diagnosis. Because Western medicine has a clear priority to find as much disease as possible, the boundaries, when they are laid, will be set to achieve that goal. The Western medical system punishes doctors who miss disease. There is also a poorly tested assumption that early detection of disease is always for the best. Scientists and doctors have every reason in the world to create over-inclusive diagnostic criteria.

There are numerous examples of how Western medicine has made patients out of people who were previously perfectly well.

For example, in 2002, a group of experts formed a committee to develop criteria that would allow the early detection of kidney disease. Renal failure is life threatening and life destroying; it is devastating to the individual and costly to the health service. The committee assumed that, by creating criteria inclusive enough to detect the earliest signs of kidney disease, they would ultimately be able to reduce the number of cases of kidney failure and thus save lives. That all seems very admirable, until you see the consequences of their work. The application of their criteria meant that, almost overnight, a substantial number of people who didn't know they had a disease, and weren't necessarily looking for one, were suddenly told they had a medical problem. It was estimated that 10 per cent of people in the US and 14 per cent of people in the UK fell into the new expanded category for kidney disease. That compares with rates of less than 2 per cent before the criteria change.

A problem arose with the number of new cases of kidney disease: if they were to be believed, as many as a third of people over sixty-five could be heading towards kidney failure. But, since only one in a thousand people actually develops end-stage renal failure every year, that could not be correct. The mismatch in numbers meant that the vast majority of those newly labelled with chronic kidney disease would never have progressed to serious kidney disease if left alone. Thus, a huge population of asymptomatic people were burdened with a diagnostic label, accompanied by the need for regular check-ups and tests that they almost certainly didn't need.

There are numerous examples like this. In 1994, when the World Health Organization created new criteria for diagnosing osteoporosis, the number of people with the diagnosis doubled overnight. The new definition of osteoporosis came with a similar problem to that of kidney disease. Studies ultimately showed that 175 of the new osteoporosis patients would need to be treated for a full three years to prevent a single hip fracture.

Oversensitive criteria created over-diagnosis of both disease groups, and a great many people faced monitoring and treatment that they probably didn't need. A very large number of healthy people had become patients through nothing more than an expanded disease definition. Most of those people will not have been aware that an arbitrary and overgenerous shift in a dividing line was enough to make them sick. They put trust in the medical community and in science, not knowing there was so much ambiguity in medicine.

Obviously, over-diagnosis creates the inconvenience and risk of unnecessary treatment, but, for the sake of people like Sienna, and others with functional and psychosomatic illness, what concerns me are the psychological and behavioural effects on those people drawn into the expanded categories for chronic kidney disease and osteoporosis. I suspect that many were grateful. The news alerted them to be more mindful about health matters, and the vast majority neither developed kidney failure nor suffered a fractured hip. They may have attributed that success to the prevention programme, which is, of course, the beauty of a very inclusive preventative programme – if you treat too many people and almost none of them develop the disease in question, you can hail it as a success. The patients will probably even praise you, not realizing they never really needed all those doctors' visits, blood tests, scans and supplements.

But a negative consequence of over-diagnosis that is hard to measure is how people embody their new identity as a sick person. Just as Hacking described, the embodiment of illness labels by the classification effect makes new people, and, if the label is powerful enough, it will make new disabled people. I wonder how many of those who were told they had kidney disease became excessively concerned about their health, searched their bodies for symptoms and limited their activities. It's very hard to extricate somebody from a label once they have inhabited it. People regard medical diagnoses, especially those based on scientific-looking test results, as infallible truths, when, in fact, they are only too fallible.

Aside from Western medicine's innate fear of missing disease, there are also personal characteristics of doctors and scientists, and practical reasons belonging to the institutions, that foster a climate of over-diagnosis. The committees who define what constitutes illness and disease are made up of small groups of people – and it is inevitable that they will have their own agendas and that they will have flaws, because we all do. When setting the limits of disease, it is logical for doctors to create parameters that bring as many patients into the fold of their own medical field as possible. Doctors are well within their rights to mine grey areas for patients – and many do.

Let me use an entirely fictional scenario to demonstrate this. If I ran a service treating people for insomnia and there were not enough people complaining of insomnia to fill my clinic, then my service would be under threat. Patients bring money into hospitals, and without them services could lose funding and staff. In that situation, expanding the definition of insomnia and then advertising it to the general public would attract new referrals. Researchers in the field of sleep who were in need of more people to study could do the same. Pharmaceutical companies selling sleep remedies and manufacturers of equipment used to investigate sleep disturbance need worried sick people to stay in business. I emphasize that I know of no such thing happening in the field of sleep research and treatment – there are more than enough people troubled by sleep problems to keep those experts busy for a very long time. However, both consciously and unconsciously, these drives to attract patients into a service are alive in many other medical fields. There are more reasons for medical specialties to detect abnormalities than to reassure people that they are normal.

The seventeenth-century physician Thomas Sydenham, sometimes referred to as the English Hippocrates, said that a disease was something waiting to be found out, which existed independent of the observer. It is a cancer that grows and makes itself known, rendering the person sick whether the doctor defines it as a disease or not. Illness is another matter. It is a perception of how one feels and does not need to be associated with a disease – i.e., it does not need an objective pathology to exist. Illness is defined by the person who has it and the doctor who gives it a name, and as such it will be an inherently cultural phenomenon. 'Normal' depends on the community in which you live. So, being overweight in Los Angeles will look very different from being overweight in Samoa. The normal range for haemoglobin level may be different for a community that lives at high altitude than it is for one that lives at sea level. We are measured against others in our society. Being healthy may just mean being healthier than the people around you. This process of defining abnormality is made by doctors and scientists embedded in a community and influenced by its concerns and values.

Apart from broadening diagnostic criteria for disease, another way to medicalize is to redefine what is considered physiologically normal, thus creating a brand new illness without the need to prove any pathology. New medical diagnoses are coined regularly, but there is a difference between those that are objective, and therefore devoid of cultural influence, and those that are sociocultural. The objective sort arise when there is a genuine novel discovery of a pathology, such as a new virus or a genetic abnormality, that explains a previously inexplicable medical problem. These diseases, as Sydenham described, would have existed and caused ill health and death, even if a scientist never uncovered the precise cause. However, there are also new subjective diagnoses, which come from nothing more than a decision to redefine the parameters of normal physiology and behaviour. These are medical conditions that do not require any proof of pathology; a doctor only has to decide that a normal characteristic has been stretched to the point of seeming abnormal.

Let me illustrate, with a very trivial example, how easy it is for a doctor to create an illness out of something that was previously considered unremarkable. In 1985, Dr G. D. Shukla published a paper in the *British Journal of Psychiatry* suggesting that not sneezing – asneezia – was an under-recognized sign of psychiatric illness. Dr Shukla, a psychiatrist, identified this symptom in 2.6 per cent of his patients. Since then, a smattering of other mentions of asneezia have appeared in the literature, although thankfully the act of not sneezing has not really taken off as a legitimate sign of disease. But it could have. Had Dr Shukla been more powerful, had his paper been interesting enough to catch the attention of the mass media, had there been a drug company selling a remedy to asneezia, then things could have been different. There have been many other much more successful novel medical problems based on a redefining of 'normal' that have taken off much more successfully. Sitting in clinic on any average day, I meet a handful of people who have a formal medical diagnosis given to some feature of their person that would probably have been called normal when I first qualified as a doctor.

I feared that Sienna had fallen prey to Western medicine's over-enthusiastic love of diagnostic classification. Of the many labels she had been given, all were the type of diagnosis made in the absence of pathology, just by setting limits on normal physiology. These are the sorts of medical problems most available to over-diagnosis.

Within a normal distribution, there will be a large number of people who are clearly normal, a much smaller number who are clearly abnormal, and a third group who lie around the border zone – and it is these border-zone people who are far too easily drawn into Western medicine's bias for disease. For example, most people have read or heard that seven to eight hours sleep per night is considered optimum. For many, that range of sleep has been taken as an absolute, as if getting less or more than that has to mean there is something wrong. But, actually, while a normal distribution is a good measure for assessing whole populations, for individuals it is only meaningful when their personal characteristics are taken into account. Therefore, when it comes to sleep, there will be a reasonable number of people who get less than the average amount of sleep, and these non-average people will still be perfectly healthy.

For these sorts of reasons, I had concerns about Sienna's diagnosis of PoTS and joint hypermobility, since both diagnoses are open to potential over-diagnosis and neither had helped alleviate Sienna's symptoms – the opposite, in fact: she had embodied the labels and her symptoms had increased.

PoTS is a disorder that rarely has objective pathology to prove it exists. In its severe form – in those patients well outside the normal range and therefore unequivocally sick – it has been attributed to a fault in the autonomic nervous system or connective tissue. However, it is the grey zone of the milder form that concerns me here, because those cases have no proven pathology. At the mild end of the spectrum, PoTS is a clinical diagnosis, without evidence of disease, but assumed to be a disease nonetheless. The development of PoTS as a new diagnosis depended on the informed, albeit estimated, placing of limits on the heart rate when standing. For all the reasons described before, that leaves it open to over-inclusive diagnostic criteria. People at the fringes of the normal range are more likely to be told they have PoTS than to be reassured they do not.

The more I learned about Sienna's diagnosis of PoTS, the more uncertain I was about the validity of that diagnosis. It had come about after she fainted. Fainting is not terribly unusual in young women, but it had concerned her enough to go to the casualty department. There, a non-specialist junior doctor suggested she be tested for PoTS. A more senior doctor or her GP would probably have just reassured her, but junior doctors order more tests because they are inexperienced. The faint caused a degree of anxiety in Sienna and she began to notice she was dizzy.

The standard test for PoTS is a tilt-table test. Head-up tilt-table testing involves holding a person upright on a tilting table while monitoring for heart rate and blood pressure changes. A heart rate rise of thirty beats per minute has been chosen as the cut-off point between normal and abnormal – so, people with a heart rate rise of twenty-nine on tilt-table testing do not have PoTS, while those whose heart rate rises by thirty beats do.

Sienna's first tilt-table test was negative, meaning she did not have PoTS. Part of Sienna's problem with medical diagnoses was that she was reassured by having explanations for every bodily symptom, so the normal result from the tilt-table test didn't sit well with her. She wanted to know why she fainted. Concerned that something had been missed, she sought out another test and this time it was positive. People stop looking for answers when they find one they can relate to, so Sienna took the second test as the correct one and accepted the label of PoTS. She could just as easily have chosen the first test as the correct one, and then maybe her outcome would have been different.

Knowing she had PoTS brought Sienna's body into her awareness. She looked up the common manifestations of her new medical problem and unconsciously enacted them. She noticed dizziness to a much greater degree and became aware of her heart rate. She began to pick symptoms out of the white noise and to worry about them. A template in her brain told her that standing could lead to collapse. Predictive coding and prediction errors began to confuse her nervous system. She began to avoid standing because of how unpleasant it felt. As internal and external feedback loops combined, her body became deconditioned, increasing her symptoms. She inadvertently trained her body into coping less well with standing. In return, the increased symptoms increased her fear, and so the cycle went on.

The concept of PoTS was first described many decades ago, but the diagnosis only entered mainstream medicine in the 1990s. It is now an incredibly common diagnosis in young women. A condition that didn't exist thirty years ago is now said by some to affect one in a hundred people in the US. What happened to those people before the PoTS diagnosis existed? I strongly suspect that, thirty years ago, Sienna would simply have been told she had a tendency to faint. She would have been given exactly the same treatment advice that she was given for the treatment of PoTS – stay hydrated, eat a high-salt diet and stand up slowly – but without the label. Not offering a diagnostic label does not mean inaction or dismissal. When you avoid a label, you avoid offering somebody a sick role, with all the negative consequences that can come with it. I suspect that – for Sienna – this strategy would have been considerably more successful.

Sienna also had a diagnosis of joint hypermobility, which was applied when she complained of joint pains in her mid-teens. Extreme joint hypermobility associated with an objective genetic abnormality has always been a disease, but in very recent years the general flexibility of a person's joints has had disease parameters assigned to it without any disease needing to be present. This has drawn a large population of 'mild' cases of joint hypermobility into the fold – people who are deemed to be too flexible to be normal, even though they do not have the genetic abnormality that goes with the severe form of this problem.

Once the concept of milder forms of hypermobility and PoTS was created, doctors started using these new definitions to give succour to people desperate to have labels for bodily experiences. As a consequence, joint hypermobility and PoTS are appearing in burgeoning numbers, and I suggest that is through nothing more than the classification effect. The diagnoses are not a problem for those who respond to them by making positive changes that lead to an improved quality of life. But they could be a big problem for many young people with a tendency to embody medical concepts in a way that promotes disability instead of alleviating it.

With the effect of labelling in mind, it is equally valid to question my diagnosis of dissociative seizures in Sienna. Dissociative seizures also have a severe form, manifesting as prolonged convulsions that nobody could mistake for 'normal'. But Sienna's milder complaints might have been more successfully addressed using a less medical explanation. In fact, a year later, Sienna experienced the disadvantages of being given such a label, and I was forced to revisit my guilt about choosing to give it to her.

Sienna had a series of meetings with the psychologist before we met again. During a course of treatment, she had improved considerably. However, it wasn't sustained. As soon as she stopped seeing the psychologist, blank spells and difficulty concentrating recurred and then got worse. Over time, she developed convulsions, which convinced her once again that she had epilepsy.

'I think I need to be tested again,' she told me. 'Maybe you didn't do enough tests the first time.'

I could not have been more comprehensive the first time, I thought, but said, 'I am absolutely sure that you did not have epilepsy the first time, and there is no reason for you to have epilepsy this time. The seizures you have described are nothing like epilepsy.' I had seen a video of her attacks and they lasted too long and the movements were of the wrong quality for epileptic seizures. What's more, she already had a confirmed diagnosis causing her blank spells, so it was counter-intuitive to think she now had something new.

'Can I have the tests again? Maybe there's a different type of scan I can have? Is there?' she said, hopefully.

There are so many ways for Western medicine to waste people's time and turn them into patients, and over-investigating with a plethora of unnecessary tests is one such way. Tests are not inert things and can do more harm than good. In fact, on my visit to Le Roy, New York, I had stumbled on the oddest reminder of this. I found it in the Jell-O museum. Located in an old brick schoolhouse on Main Street, the museum commemorates the region's most famous export. It is one of those places that you do not expect to find all that interesting, but which turns out to be oddly fascinating.

During my personalized guided tour, which the ladies who work at the museum offer to every visitor, I had not expected to find one of medicine's cautionary tales. However, taking pride of place in the museum was an EEG of a Jell-O, with an accompanying newspaper headline which read, *EEG proves your brain isn't mush, it's Jell-O.* In 1974, an Ontario-based neurologist called Adrian Upton did an experiment in which he attached EEG electrodes to a dome of lime jelly and made a recording. He likened the resulting tracing to the brainwaves of a human and suggested that this was potential evidence of life. The EEG recording of the Jell-O is on display in the museum alongside a human EEG, and the blurb attached states that 'lime Jell-O has qualities virtually identical to the brains of healthy men and women.' Of course, this isn't true. To the uninitiated, the two EEG recordings may look similar, but to a neurologist the two tracings are wildly different. Dr Upton was not being whimsical when he carried out this study – he aimed to make a very important point. Medical tests are not harmless. Many, including the EEG, are fraught with a high potential for false positives. The results are open to interpretation and must be taken in a clinical context. This applies to a huge number of medical tests. Do enough of them and a doctor will always find some variance from normal that is easy to over-interpret, adding to patient anxiety levels. More and more tests do not equate to good medicine. Appropriate medical tests, done by people who are experienced enough to properly interpret the results, are good medicine. Following this basic principle is important if a doctor wants to avoid finding themselves resuscitating a jelly.

I tried to dissuade Sienna from taking the route of scans and EEGs, which I was certain would lead her nowhere. She argued that she could not have dissociative seizures because, if she did, the time she had spent with the psychologist would have cured her. That only reminded me again that functional disorders are held to higher standards than any other type of diagnosis. If a person is being treated for migraine or epilepsy or diabetes and they don't get better, they usually ask for a different treatment, whereas the failure of treatment for functional disorders causes people to ask for a different diagnosis.

I spent another six months reinvestigating Sienna, only to get exactly the same result: she did not have epilepsy and her convulsions were caused by dissociation. She met the psychologist again, but did not get better, because not everyone does. She ultimately dropped out of university. Looking back, I think the only point at which her trajectory could have been slowed was with the first of her medical diagnoses, not the fifth. From the age of fifteen, she had been taught to medicalize every bodily change and, once that pattern was set in her brain, she struggled to escape it.

Western medicine's love of drawing people into diagnostic categories and applying disease names to small differences and minor bodily changes is not specific to functional disorders – it is a general trend. Pre-diabetes, polycystic ovaries, some cancers and many more conditions have all been subject to the problem of over-inclusive diagnosis. My biggest concern in this regard is the degree to which many people are wholly unaware of the subjective nature of the medical classification of disease. If a person is told they have this or that disorder, they assume it must be right. The Latin names we give to things and the shiny scanning machines make it look as if there is more authority than actually exists. To a certain extent, Sienna pursued each diagnosis she was given, but other people have diagnoses thrust upon them, having no idea that there might be anything controversial about it – and having no idea that they have a choice.

Western medicine's hold on people, and its sense of being systematic and accurate, makes it a powerful force in the transmission of cultural concepts of what constitutes wellness or ill health. But Western medicine is just as enslaved to fads and trends as any other tradition of medicine. Much of today's science will not be considered science in fifty years. In 1949, the Portuguese neurologist António Enas Moniz won the Nobel Prize for Medicine for developing a technique to do a frontal lobotomy, which was used as a therapy for psychosis. Unfortunately – for women, in particular – the limit of what constituted 'normal behaviour' was very narrowly defined at the time, which led to the procedure being used to control rebellious daughters and cheating wives. Disease definitions change all the time, sometimes to our benefit, but not always.

I would not wish to put the problem of over-diagnosis wholly at doctors' doors. Most of us are working very hard to act as the gatekeepers, trying to prevent overenthusiastic diagnosis, trying to reassure people that they do not need tests and that their health is good. But this simply doesn't work as well as it should. Increasingly, people demand labels and tests. Society pressurizes doctors into giving absolute answers and punishes us for missing things. The accumulation of new and expanded medical categories is a collusion between doctor and patient. Some have implied there may be a darker reason why doctors like to diagnose, thinking it a means of social control. Giving something a medical label makes it a doctor's business. I still truly believe that most, although certainly not all, scientists and doctors are working for the greater good – but we are not always as clever as we think we are, and are certainly slow to learn from our mistakes. The opioid crisis and the problem of antibiotic overuse leading to antibiotic resistance do not seem to have done anything to dampen our enthusiasm for over-medicalization. In the field of functional neurological disorders, doctors see the end result of embodied disease labels and it can be a grim sight. Sienna is now in her late twenties and has so many chronic medical problems, all without proven pathology, that she will never be able to live in the world normally. I see people almost exactly like her, several times a month.

If it is that difficult to decide on the absence or presence of disease, just imagine how hard it must be to define what constitutes normal behaviour. If a rebellious teen was pathologized seventy years ago, what personal traits are we attributing to illness now, that we will regret in the future?

The *Diagnostic and Statistical Manual of Mental Disorders* (*DSM*), created by the American Psychiatric Association, is on its fifth edition. With every new edition, it grows, as new categories of psychiatric and psychological illness are added. What constitutes a mental illness changes substantially over time. Some classifications disappear, but the trend is towards broadening categories with new definitions and subcategories. The *DSM-IV* excluded grief from the diagnosis of severe depression, but the *DSM-V* does not make that exclusion. Some argue that this risks medicalizing ordinary grief, but others say omitting grief could deprive a group of people of getting the help they need. There is no single right answer.

Much like the definition of kidney disease or osteoporosis, the *DSM* is produced by a committee of experts, and as such it is a culture-bound document, tied to a time and place. That is best illustrated by the fact that the *DSM-I* listed homosexuality as a sexual perversion. In the *DSM-II* and *-III*, it became a sexual-orientation disturbance. It only lost its classification as a mental illness, and was subsequently dropped from the *DSM*, in 1973.

Included in the *DSM* are neurodevelopmental disorders, like autism and attention deficit hyperactivity disorder (ADHD).

These diagnoses are made with increasing frequency. When I was a junior doctor, these labels were mostly given to children, but now they are increasingly diagnosed in adults (criteria suggest the features should always have started in childhood, although not every diagnostician adheres to these strict diagnostic criteria). The steep rise in diagnoses is a good thing, for the most part; there is no doubt that ADHD and autism can be serious, life-limiting conditions, and that the children affected by them used to be neglected. The increased prevalence reflects the fact that we now actively look for children who need extra help, which allows us to provide the support needed to circumvent educational problems that, in the past, might have seen them dropping out of education altogether. Recognizing specific deficits has stopped children from being labelled as stupid or badly behaved, and has allowed them work to their strengths.

But there is a flip side. The rise in numbers is partly due to the enthusiasm with which these labels are being sought out and applied. Here, again, we have the problem of potential overdiagnosis looming large, with not nearly enough consideration of the drawbacks that might bring. Normal child development is subject to social judgement. It differs across cultures and is not an absolute. Because there is no single diagnostic test for problems like autism and ADHD, and because the criteria are qualitative, that leaves them open to a certain degree of interpretation by the user. With terms such as 'doesn't seem to listen when spoken to', 'forgetful in daily activities', 'fails to finish homework', 'talks excessively' and 'fidgets' used in the diagnosis of ADHD, one can see how much leeway there is for the diagnostician. How much restless, fidgety, inattentive behaviour is too much? These features have to persist for more than six months and must interfere with function, but that depends on what you consider acceptable normal function to be. The diagnosis is contextual: a person is measured relative to others in a social group. Diagnostic criteria can be applied differently by different people.

The prevalence of ADHD rose steadily in the US, from 6.9 per cent in 1997 to 9.5 per cent in 2007, with further rises since. In his book *ADHD Nation: Children, Doctors, Big Pharma, and the Making of an Epidemic*, Alan Schwarz draws attention to much higher rates in some states, with Virginia coming in at 33 per cent. One study showed that more than 20 per cent of high-school boys had been told they had ADHD. There are a lot of controversies in the field. In Iceland, a survey showed that children born in the second half of the academic year, who are therefore the younger ones in a class, are more likely to get a diagnosis of ADHD, which raises the possibility that we do not know the difference between immaturity and ADHD. There are also cultural differences – for example, Hong Kong has particularly high rates, which some suggest means that anger and strong emotion are more likely to be pathologized in Chinese culture.

There are numerous factors driving over-diagnosis, and commercialism is undoubtedly one of them. Pharmaceutical companies must have made billions from the increased use of methylphenidate (Ritalin) to medicate adults and children who are diagnosed with ADHD. Ritalin is even used to enhance academic performance in people without a medical diagnosis, although it is easier to access the drug if they are medically confirmed as having ADHD. The subjectivity of the diagnostic criteria and shifting definitions have also contributed to the rise in cases. Worryingly, there are many incentives pushing parents, teachers and schools into pursuing one of these labels. If a child gets extra help or a school gets more resources, then it may be in everybody's interest to be as inclusive as possible when a diagnosis is made.

It is very important that these sorts of problems are recognized in children. I should again emphasize that my concern applies to the effect of labelling in the mild spectrum, not to the diagnosis as a whole. Anybody who has seen a child with severe autism or ADHD would not contest the level of their disability. My concern is about the long-term harm to those children in the grey zone of the diagnosis, around the edges, where there are harms that are not always given enough consideration. A child labelled in this way may be viewed differently, assumed to be less clever and less likely to succeed. The child may identify with the label and fulfil the prophecy associated with it, a stigma they may then carry for their whole life. What's more, over-diagnosis of adults and children based on a poor concentration span risks trivializing the difficulties of those with severe forms of this problem.

It may be that children identified in school as having mild ADHD have been correctly identified as needing extra help, but surely a better solution would be to provide that help without making a medical diagnosis. It is the label that pathologizes behaviour. Do we need it in order to pay attention to children who are struggling in school, or to notice a child who is failing to make friends?

The diagnosis of learning and neurodevelopment problems in adults is also common in the twenty-first century. Western society values resilience, achievement, independence and individual success. If a person fails in something they really want to do, society is more likely to advise them to keep going than it is to suggest they change their ambition: 'If at first you don't succeed, try, try again.' We encourage our children to never give up on their dreams. We tell people that if they really want something, they will eventually make it happen. Writers who are struggling to be published are inevitably told about J. K. Rowling's multiple rejections by publishers. The moral of the story is that persistence pays off. It's not true, of course – it doesn't work out for everybody. And, even when it does, the effort of achieving the goal can be too much for the mind and body, and can make people sick.

Sometimes, illness is a sign that the life we have chosen for ourselves is not the right one, but Western culture doesn't make it easy to acknowledge that. There is an increasing tendency for people to seek a medical reason to explain why things are not working out. I suspected that Sienna's dissociation was her body's way of telling her that her life choices were not suited to her. Westernized values put people at risk of judging themselves to be failures if they do not meet society's measure of success. More and more people are seen to lack – or believe they lack – what they perceive to be the right personal characteristics, and they then seek medical explanations to account for it.

Two stories I heard on the radio come to mind. In the first, a woman was talking about being diagnosed with autism in adult life. She had been depressed for years and had struggled at work. When she was finally assessed by a psychologist and was given the diagnosis of autism, it had helped her re-evaluate her life.

'It was a huge relief,' she said. 'The things I thought were negative aspects of myself, and failures and flaws, and things I couldn't control, were suddenly put into a context where they made sense . . . Work was a huge emotional and physical stress . . . so I quit my job very soon after and went off and pursued a very different career. I wouldn't have known to do that had I not had the diagnosis.'

Why did she need the diagnosis? I wondered. She was

struggling and unhappy – shouldn't that have been enough to tell her to reassess her life? Being autistic gave her permission to leave a job she hated. That seemed to me to be too complicated a route to get to a decision that should have been obvious. It made me think of Lyubov and Krasnogorsk, and all that had to happen there before the townspeople felt able to give up the town they loved.

In the second story, an author was talking about the book she wrote about her mother's exceptionally late diagnosis of autism at the age of seventy-two. The author described her mother as depressed and anxious throughout her life. The mother repeatedly told her family that she felt dreadful and that life had no meaning. She also had multiple physical complaints, necessitating lots of tests, none of which led anywhere. Doctors subjected her to many invasive investigations and then discharged her without a diagnosis. In the end, even her own children tired of it. They stopped listening.

Then, very late in the mother's life, a doctor diagnosed her with autism, and it was a turning point for the author. It completely changed how she saw her mother's suffering and awakened a new compassion for her. But why were her mother's protestations that she was so desperately depressed not enough? Surely a statement that life has no meaning is as clear as it can get? Why could neither family nor doctors read the need in her repeated presentations to hospital with inexplicable physical symptoms? Westernized society simply has less time for psychological suffering than for other sorts. Our brains shut out that sort of complaint and thus we drive people to get a disease label that will earn them the help and respect they are asking for.

Specialist doctors in Western medicine also take symptoms very literally. Family and community doctors are better at seeing the whole picture, but hospital-based doctors can investigate single symptoms in an entirely non-holistic way, and then discharge their patient without help, if their tests are normal. Hwa-byung, grisi siknis and resignation syndrome are culturally specific languages of distress. The symptoms have meaning beyond the organs involved. A chest pain doesn't necessarily mean a heart problem. Western doctors have trouble tuning in, or at least in knowing how to respond, when something in the picture doesn't fit the disease pattern they are trained to treat.

We in the West also live in a culture that prizes happiness so highly that anything less risks being classed as abnormal. We medicalize and commercialize human suffering. Anglo-American cultures tend to conceptualize depression as physiological and psychological, whereas other cultures regard it as situational. Again, the controversy in the diagnosis of depression lies around the peripheries. Severe depression has fairly stable symptomatology: it obviously affects a person's ability to function and clearly demands that they receive help. But mild depression is a more disparate category, where the border between unhappiness and illness is harder to define. Western cultures push towards illness. That would be a positive thing if it made people better, of course, but not if it is merely a label that creates chronic illness.

In a 2013 paper called 'Depression as a Culture-Bound Syndrome', the GP Chris Dowrick considers the factors that drive the medicalization of mild depression. There is a clear commercial and practical element. A great deal of money is made from the treatment of depression and, for the purpose of insurance companies and doctors' visits, it helps to have a category into which to fit consultations for low mood. Doctors like to offer a diagnosis and, for patients, it gives meaning to their feelings. However, as Dowrick points out, the utility of the diagnosis is debatable. While antidepressants have been shown to provide benefit to people with severe depression, meta-analysis of clinical trials provide little evidence for their use in mild depression. In trials, their use in mild depression is only as good as a placebo. Dowrick argues, and I agree, that labels make patients passive victims of circumstance. To reconceptualize depression and low mood outside of the frame of these labels doesn't have to mean that a person cannot turn to a doctor for support. It would seem preferable to offer a person a management role over their own distress, rather than a role as a passive sufferer.

In writing about the cultural variation in the clinical presentation of anxiety and depression, the psychiatrist Laurence Kirmayer points out that the vast majority of people in the world do not consider depression to be psychiatric, and therefore conventional psychiatric approaches cannot possibly help them. In the UK, I have often heard people refer to their own low mood in terms of low serotonin levels, but evidence shows that ethnic minorities are more likely to regard the same sort of feelings as a consequence of life events, rather than being physiological or related to their mental health. Medical papers suggest a keenness on the part of Western medicine to draw people into our classification system, referring to mental-health issues as under-recognized in minority groups. The *DSM* is a culture-bound document, but the medical establishment is keen to apply our rules to other people because we assume our evidence-based, scientific approach to be superior. The problem is that the evidence base does not represent the majority of the world's population. The bulk of mental-health research is done on white, educated people, living in industrialized, wealthy countries. The groups on which we are keen to impress our system of medicalization are not represented.

It all makes one wonder how the Korean hwa-byung sufferers are faring in their adopted nation of the US – they will inevitably be assessed and treated according to the Western medical tradition, because modern medicine is not good at incorporating other viewpoints. They will be referred for medical tests or counselling or both. All the subtlety of the hwa-byung language will be lost on doctors whose viewpoint on suffering is first and foremost biological, only turning to psychology when the biological theory fails, giving them little capacity to incorporate social and cultural factors into their work.

In his book *Crazy Like Us*, Ethan Waters questions the certainty that Western cultures have in their approach to mental-health issues. He suggests that we invest in drugs and talking therapies for depression because we have lost the sense of community that used to provide support for struggling people. He uses the example of trauma counselling and describes the attitudes of US psychologists who visited Sri Lanka in the aftermath of the 2004 tsunami. In the months after the disaster, individuals and groups of US volunteers travelled to Sri Lanka with the aim of both providing support and training local people in the type of medical treatment they thought the Sri Lankan people needed. It did not always go well. When trying to instruct local people on how to support the victims, psychologists referred to them as being poor listeners who were not very 'psychologically minded'. The psychologists brought the concept of PTSD to Sri Lanka. They viewed the reaction to stress as something physiological happening in the brain that was universally the same, rather than being intimately linked to society and culture. They assumed that their type of counselling was always the right approach and made other assumptions about the Sri Lankan people. They viewed them as vulnerable as a result of years of poverty and war. The US clinicians were confused by Sri Lankan behaviour in response to the disaster – for example, judging the children's wish to return to school immediately as denial. Gaithri Fernando, also a US-based psychologist, but Sri Lankan born, watched with unease as Western PTSD researchers and counsellors flooded the country. She regarded the hardship as having made the people resilient rather than vulnerable. She knew that the Sri Lankan way of experiencing stress was different from the North American way – the Sri Lankans were more likely to express it as physical symptoms and lacked the mind–body disconnect that still lingers in Western medicine. They were also more likely to experience the trauma in terms of the effect it had on social relationships, rather than just as an internal state. She emphasized the role of spirituality in helping people to cope – an aspect that the Western physiological and psychological approach neglected. In fact, Fernando had previously studied the effect of war on the community and had discovered that Buddhist and Hindu children seemed to be less likely to suffer depression than Christian children. The US treatment pathway, meanwhile, was sorely lacking in cultural sensitivity.

The point, here, is not to say that one culture has a better system than another for approaching mental-health problems, but to say that that there is no one right way and it is therefore ill-advised for one community to think it knows what is right for another. Or, for that matter, for one person to impose their way on another. In describing cultural differences in the approach to grief, Laurence Kirmayer gives an example from Latvian culture in which the grieving person is advised to 'bury their suffering under a stone and step over it singing'. Australian Aboriginal people feel a deep connection to the land. They cannot be healthy unless the land is healthy. Western medicine doesn't care about the Latvian or Aboriginal perspective. It may work for us, or maybe it won't, but it shouldn't be assumed to be appropriate for everybody.

Conventional modern medicine is powerful and is always keen to recruit others into its way of thinking and practising. Evidence that a problem is medical rather than situational is often drawn from the fact that medication can change the problem. So, if antidepressants make you feel better, does that confirm you were depressed? If Ritalin improves your concentration, is that evidence of ADHD? The fact that a person can optimize their abilities with medication doesn't provide evidence of illness. If it did, athletes taking performance-enhancing drugs could use it as proof that they have a medical disorder that holds them back. Science offers us the opportunity to conceptualize failure, loss, grief and sadness through illness and disease, and many of us take up that offer. It makes sense for us to medicalize low mood or personal flaws, because diagnosis helps us to escape being defined as lacking in the characteristics that are valued by our culture. But perhaps a better answer is to provide more support to people, so they can reassess their situation without the requirement of a formal medical diagnosis.

Our brains hate chaos and are soothed by answers. A medical diagnosis that seems to make sense – the more scientific sounding, the better – goes a long way to ease the pain of life. I'm not necessarily saying that is wrong. If it helps a person feel better and move forward, then it's a good thing and I support it. I no more object to these Western formulations than I do to the beliefs of the Miskito and Wapishana people. Neither medicalization nor spirituality are coping methods I choose for myself. As a doctor, I am only interested in what makes a person feel better. But the choice should be made with all the information available. A medicalized route should never be foisted on a person by the medical establishment, any more than faith should be forced on us by the Church. If the medical establishment peddles scientific superiority and certainty as equal for every sort of diagnosis, then that removes informed choice.

I am most concerned for future generations, however. Adults can go to a doctor and seek out these labels for themselves. They can keep or reject them, depending on the impact they have on their quality of life. But, in turning our attention to our children, we risk putting them at the constant mercy of newly expanded and created diagnostic categories that try to explain their weaknesses through learning and socialization disorders and physical disease. These labels are offered with insincere certitude. In a medical climate of heavy over-medicalization, the diagnoses we give children are not reliable, but parents think they are. Who knows how those labels will affect the children psychologically and practically when they get older? The *duende* comes and goes, but diagnoses like autism, ADHD, depression and PoTS are forever.

Epilogue

Society: A body of individuals living as members of a community.

In 2018, resignation syndrome moved to Nauru, an island in the Pacific. At the time, Nauru was being used as a holding pen for refugees seeking asylum in Australia. True to the form of biopsychosocial disorders, the symptoms of resignation syndrome changed when it arrived in another continent. Here, children refused to eat. They expressed outright depression and desperation before they stopped moving. The Swedish children had slipped into a listless, apathetic world – they were passive, but the Nauru children's condition was an active cry for help. One young girl set herself on fire. Children begged to die. The medical condition had evolved to meet a more immediate, urgent need.

Nauru is a tiny island. Its natural resources were destroyed by mining long ago, and its marine life culled by pollution. The children interned there had no way to move forward in their lives and had no idea where their futures lay. Their island home was more like a prison, complete with guards. Their chances of being admitted to Australia were slim. They could be sent back to the country they came from or offered asylum somewhere far away. As awful as the plight of the Swedish children clearly was, in Nauru, it was much, much worse. By 2018, resignation syndrome had been widely reported in the world's press, so the spread to a related group half a world away was not surprising. In 2019, it spread again to the refugee camps on the Greek island of Lesbos.

The question arises, what should be done to prevent future cases of resignation syndrome and how should those already comatose or catatonic be treated? When I visited Nola, Helan, Flora and Kezia, they were lying inertly in their homes, with no active intervention taking place. If this is a medical illness, then surely the children should be in hospital or, at the very least, they should be receiving intensive treatment at home. If they were European born – if they were not refugees – society would expect a lot more for them.

Or should the question be, is theirs even a medical illness? If it is a sociocultural phenomenon, then perhaps it is just about tolerable that the children wait at home, while people like Dr Olssen call for a more definitive solution for them. If they were hospitalized and treated individually, they might stand a fighting chance of recovery, but where would that leave the underlying social problems that have created the disorder? Mario, my new Miskito friend, speculated that disorders like grisi siknis made girls strong who were otherwise weak. I came to regard resignation syndrome as a language of distress that speaks louder than words. Would treating the symptoms of resignation syndrome without addressing the root cause ultimately rob the community of their voice and the children of their strength? If resignation syndrome was abolished, how would the message about the plight of asylum-seeking children get out? Clear statements of distress and standard calls for help do not seem to be enough.

As I watch resignation syndrome advance around the world, and as I hear of new outbreaks of psychosomatic illness taking hold of brand new groups of people, I despair of the havoc they wreak and how poorly understood they are. When I started writing this book, I had a naive hope that offering people an understanding of the types of physiological mechanisms thought to create functional disability would somehow help protect against them. Then I started listening to people's stories – for once, not as their doctor, and freed of the obligation to cure them. As a result, I found a new perspective that made me wonder if eradicating these disorders was, in fact, the wrong thing to hope for. For many of the people I met, psychosomatic illness served a vital purpose. Seizures solved a sociocultural problem for the Miskito and Wapishana people. In Krasnogorsk, a sleeping sickness did the same. Psychosomatic and functional disorders break the rules of every other medical problem because, for all the harm they do, they are sometimes indispensable.

There are simply not enough words to express everything a person feels. The complexity of human emotions cannot be distilled into something rational and well thought-out for every person in every situation. Cognitive dissonance exists, as do moral dilemmas, inconceivable choices, inequality and despair. Life will always find a way to set traps that seem impossible to escape. People are not machines, making decisions from algorithms, logical and free of emotion, so perhaps we need release valves and coping mechanisms, face-saving ways of addressing conflict and grappling with ambivalence. Sometimes, embodying and enacting conflict is either more manageable or more practical than articulating it.

When Lyubov told me about her sleeping sickness and about her life in Krasnogorsk, I saw her circumstances, and those of her neighbours, as unique. Yet, the more I reflect on it, the more her story seems to be universal, a story of lost love. What happened to Lyubov was not that different from dealing with the final dissolution of a once-loving relationship that has worn itself out. Many people have felt the extreme uncertainty and sadness of being in a failing relationship, where the question of whether to leave or not to leave has a different answer every day. Such a dilemma has a practical, analytical side that can be reduced to sensible lists of pros and cons, but it also has a much more complex emotional side that is altogether harder to unpick. The same can be said for many major life changes. Where decisions and problems are too overwhelming and close to our hearts to be fully reconciled, we may need unconscious processes to help us get through the mire. On the one hand, one could regard Lyubov's sleeping sickness as a dysfunctional way of coping with a difficult choice. Or one could regard it as the road she had to take to make a difficult decision and to deal with a great loss.

My mind also returns to the woman I heard talking on the radio about the positive change that an autism diagnosis brought to her life. For me, her story raised concerns about the excessive use of labelling and the harm of medicalization, but maybe I just failed to see the beauty of what she achieved. If I were her doctor, what would be the value in attempting to dismantle the route she had found to make her life a happier one? Western medicine encourages a literal approach to symptoms. Doctors expect people to be able to verbalize and intellectualize their problem when asked to do so. Slaves to methodology, protocols and technology, we are losing the vital skills required to decode our patients' stories. Only in an implausibly perfect, analytic world would it ever be possible to fully dispense with metaphors for distress.

Much of Western medicine's current attention is on the physiology of these disorders, because there is a demand to prove they are 'real' and there is a perception that this can only ever be achieved by uncovering changes on brain scans or in blood tests – the sort of proof Western society respects. A biological approach did work for Tara. When she lost the ability to walk, triggered by a slipped disc, the story she told herself about what was happening inside her body was central to the looping effect that created her disability. That story needed to be heard before it could be addressed, but it was only by realizing that aberrant physiological processes were at play that Tara was able to overcome her fear that the doctors had failed to diagnose a progressive disease. She could only get over that psychological barrier once she understood that the body is truly capable of this sort of trickery.

As with any other medical disorder, psychosomatic and functional disorders arise through physiological changes and that should never be underestimated. In the face of an onslaught of information, the brain is in a constant state of predicting, discarding, assessing and reassessing, drawing inferences and learning. Like cells that grow too fast and become cancerous, or organs that produce too many hormones, unconscious psychological processes are fallible and they get things wrong – all the time. Functional neurological disorders are the brain's coding errors. They are the neural circuits' dysfunctional response to a change in behaviour. They have numerous triggers, only some of which are related to psychological distress. They can be a response to injury, disease, false medical beliefs, hardship, conflict, contagious anxiety. Disability develops through a process akin to learning. But brains that have been programmed can be unprogrammed, so that should never be an irreversible process.

But where the Western biological approach falls short is when it comes to making people better. The embodiment of psychological conflict is a necessary fact of life, but chronic disability due to functional and psychosomatic processes should not be. Yet only 30 per cent of people suffering from dissociative seizures make a full recovery. Tara was one of the lucky ones, because the reality is that many functional neurological disorders have a similarly bleak outlook. I cannot think of many other medical problems that have advanced so little in terms of treatment and recovery rates. Perhaps the reason is that treatment is only ever aimed at various combinations of biology and psychology, while neglecting sociocultural influences. Perhaps doctors who do not find a way of incorporating the social factors that influence illness into their formulations will always find themselves pushing against the tide, their voices drowned out.

Of all the stories I was told in researching this book, the ones that ended most happily did not do so through biological or psychological treatments, nor even through medicine. The Miskito achieved recovery through ritual. The people of Krasnogorsk, invested in the poison theory, moved away from the source, thus allowing the sleeping sickness to solve the problem it had come to address. The Wapishana girls left the school that was making them sick. The young women of Le Roy did not accept the conversion-disorder diagnosis either, but they, too, recovered, by removing themselves from the media storm that was making them worse. Nonetheless, for the Le Roy girls, at least, their doctors' strong stance did have some positive impact, but they are the only group I discuss for whom this is the case.

The biological changes that create psychosomatic symptoms are similar in everyone, but it is the response at a community level that often makes the difference between recovery or descent into chronic illness. The quality of a person's experience is changed by others' reactions to it. I did not meet a single community that did not outwardly object to the medical concept of psychosomatic disorders and conflate it with 'faking'. Families and communities rejected the diagnosis for their loved ones, often giving the victims little choice in the matter. All the groups I learned about were embedded in a society that played a crucial role in contributing to the folklore of their condition. The communities created narratives to explain these illnesses, and those narratives led to recovery for some and long-term disability for others. In Cuba, the rumour of sonic weapons gained such prominence that some of the US embassy staff genuinely believed recovery was impossible. Having been labelled 'crazy', the El Carmen girls were ostracized from society, forcing them to prove they were indeed sick.

Of all the people I met, it was the Miskito community I came to admire the most. Theirs was the most elegant solution, because it brought people into a group rather than creating outcasts. Grisi siknis was a rallying cry that attracted a community response, whereas my patients often find themselves isolated by disability. Grisi siknis not only taught me the beauty of exteriorizing conflict through ritualized illness, but, more than that, it reminded me of the value of spirituality. As a hardcore atheist, inclined towards pragmatism, who comes from a country where religious institutions have far from covered themselves in glory, that was rather unexpected. But then I saw with my own eyes how their spiritual beliefs brought the Miskito people together and created a supportive environment that fostered recovery. I witnessed the comfort people got from the explanations those beliefs provided, and I felt for our individualistic Western societies, which are so often entirely lacking in good support systems for those in crisis.

When societies lose a shared spirituality and a sense of community and family, people have to find new avenues of support. If a person lives in a community where the only caring institutions are medical ones, then medicalizing social and psychological distress makes perfect sense. If the only place a person feels heard is in their doctor's office, then relinquishing medical illness would be counterproductive. Perhaps part of the reason the medical establishment has been so unsuccessful in treating these disorders is because there is such a deficiency in alternative sources of support. Many of my patients, not only those with functional neurological disorders, would prefer better social support to increased medication, but I don't have it to give to them. Sometimes they ask me, 'If I get better, do I have to stop seeing you?' It is a sobering thought that a sporadic fifteen-minute consultation with me, or with any doctor, could be a person's only lifeline.

In writing this book, I was reminded that a doctor must work with, not against, the formulations their patients use to explain illness. If the medical paradigms and flowsheets are not working, then doctors must stand back and listen to the story the symptoms are telling. The most graceful solutions arise when doctors and patients find common ground.

I also learned that the best chance of recovery comes when you surround yourself with a community that allows patients and their doctors to find that common ground. A community that can listen without judgement. A community that provides support. A community that can tolerate imperfection and failure, and which has the humility to put aside its vested interests. A community that is able to take a holistic view of health.

Now all we have to do is create it.

Acknowledgements

I owe a great deal to the people who allowed me to tell their stories in this book. Thank you to each of them for their generosity, trust, humour and kindness – and thanks of course for all the cups of coffee, tea and fabulous meals I was served along the way (special mention here to Lyubov whose hospitality was remarkable). Because of the sensitive and personal nature of their stories, some names have been changed, but I have done my best to tell every story as it was told to me. Still, it is possible there were times I misunderstood and if that has happened I hope I will be forgiven.

I am also deeply indebted to the people who facilitated my research and travel. Thank you in particular to Elisabeth Hultcrantz, Ed Paulette, Karl Sallin, Dinara Salieva, Danabek Bimenov, Sarah Topol, Courtney Stafford-Walter, Catalina Hernández, Mel Espinoza, Jamie Pablo Scott, Moses Lewis, Howard Owens, Kelly O'Connell, the Takotsubo Support Group and Alison Thompson. I felt particularly inspired by Dr Maddalena Canna, both by her work and her generosity. I learned a great deal from everybody I encountered along the way, but any errors in this book are mine alone.

While in Nur Sultan, Kazakhstan, a stranger called Mr Suleyman invited me to his lavish birthday party for no other reason than he could tell I was alone in an unfamiliar place and was longing to talk to people. Thank you to him and his lovely family for such a kind gesture and to all the other wonderful people who welcomed me during my various travels. And thanks to the member of the Cambridge University Irish Society who reminded me that it is easier to program a computer to beat a chess grandmaster than it is to create a machine that perfectly mimics the human gait. Our body is a mechanical wonder and yet we learn every sophisticated set of movements, every keystroke of the computer, every kick of the football, every new word spoken so casually – why so difficult to think these things can be unlearned?

Thanks to the incredible team at Picador. I am endlessly grateful to my editor George Morley for all her advice, guidance and hard work. The contributions from Paul Martinovic and Penelope Price have also been invaluable to me. Thanks to Gabriela Quattromini, Rebecca Lloyd, Chloe May, Stuart Wilson and to all the other hardworking publishing people I haven't met yet, but hope to.

Sincere thanks to my agent Kirsty McLachlan at Morgan Green Creatives who took a chance on me and for that I am eternally in her debt. And to Lisette Verhagan and the team at PFD for their enthusiasm for the book.

There was a point at the start of this book when I thought I would never find the way in. A mixture of thanks and apologies to anyone who had to advise and counsel during that time. They are too many to name.